Insulating Concrete Forms for Residential Design and Construction

Insulating Concrete Forms for Residential Design and Construction

Pieter A. VanderWerf, Ph.D.
Stephen J. Feige
Paula Chammas, M.S.E.E., M.S.B.E.
Lionel A. Lemay, P.E., S.E.

McGraw-Hill

New York San Francisco Washington, D.C. Auckland Bogotá
Caracas Lisbon London Madrid Mexico City Milan
Montreal New Delhi San Juan Singapore
Sydney Tokyo Toronto

Library of Congress Cataloging-in-Publication Data

Insulating concrete forms for residential design and construction /
Pieter A. VanderWerf ... [et al.].
    p.   cm.
Includes bibliographical references and index.
ISBN 0-07-067033-1
1. Concrete houses—Design and construction. 2. Insulating
concrete forms. I. VanderWerf, Pieter A.
TH4818.C6I57   1997
693'.5—dc21   97-19355
                                                                                               CIP

McGraw-Hill

A Division of The **McGraw·Hill** Companies

Copyright © 1997 by The McGraw-Hill Companies, Inc. All rights reserved. Printed in the United States of America. Except as permitted under the United States Copyright Act of 1976, no part of this publication may be reproduced or distributed in any form or by any means, or stored in a data base or retrieval system, without the prior written permission of the publisher.

1 2 3 4 5 6 7 8 9 0   FGR/FGR   9 0 1 2 0 9 8 7

ISBN 0-07-067033-1

The sponsoring editor for this book was Zoe G. Foundotos, the editing supervisor was Peggy Lamb, and the production supervisor was Pamela A. Pelton. It was set in Century Schoolbook by Ron Painter of McGraw-Hill's Professional Book Group composition unit.

Printed and bound by Quebecor/Fairfield.

 This book is printed on recycled, acid-free paper containing a minimum of 50% recycled, de-inked fiber.

McGraw-Hill books are available at special quantity discounts to use as premiums and sales promotions, or for use in corporate training programs. For more information, please write to the Director of Special Sales, McGraw-Hill, 11 West 19th Street, New York, NY 10011. Or contact your local bookstore.

> Information contained in this work has been obtained by The McGraw-Hill Companies, Inc. ("McGraw-Hill") from sources believed to be reliable. However, neither McGraw-Hill nor its authors guarantees the accuracy or completeness of any information published herein, and neither McGraw-Hill nor its authors shall be responsible for any errors, omissions, or damages arising out of use of this information. This work is published with the understanding that McGraw-Hill and its authors are supplying information, but are not attempting to render engineering or other professional services. If such services are required, the assistance of an appropriate professional should be sought.

Contents

Introduction xi
Acknowledgments xix

Part 1 Performance — 1

Chapter 1. ICF Characteristics — 3

- Cost — 3
  - Wall Construction Cost — 3
  - Important Influences on Wall Cost — 4
  - Construction Cost Savings — 6
  - Cost Trends — 6
  - Operating Costs — 7
- Energy Efficiency — 7
  - R-value — 7
  - Air Infiltration — 8
  - Thermal Mass — 9
  - Combined Effect — 10
  - Code and Regulatory Treatment — 11
- Design Versatility — 11
  - Footprint — 12
  - Openings — 12
  - Surface Relief — 13
  - Projections and Spans — 13
- Construction Logistics — 15
- Structural Capabilities — 15
- Disaster Resistance — 16
  - Fire — 16
  - Wind — 17
  - Earthquake — 17
- Moisture Movement — 17
- Insect Resistance — 18
- Durability — 20
  - Concrete — 20
  - Foam — 20
  - Foam-Concrete Interface — 21

Contents

Sound Attenuation	22
Toxicity	22
References	24

Part 2 Components — 27

Chapter 2. ICF Systems — 29

Categories of Systems	29
Formwork Components	32
Parts of Stretcher Units	37
Cavities and Concrete Sections	39
Specialty Units and Formwork Accessories	40
Corners	40
Ties	42
Lintel Blocks	44
Miscellaneous	46
Bracing and Scaffolding Systems	47
Available Systems	47
Flat Panel Systems	48
Grid Panel Systems	55
Post-and-Beam Panel Systems	55
Flat Plank Systems	57
Flat Block Systems	59
Grid Block Systems	61
Post-and-Beam Block Systems	62
References	63

Chapter 3. Plastic Foams — 65

Classification	65
Polystyrenes	67
Polyurethanes	67
EPS-Cement Composites	68
Properties	68
Polystyrenes	68
Polyurethanes	78
EPS-Cement Composites	79
References	80

Part 3 Design — 81

Chapter 4. Detailing Considerations — 83

Fire	83
Structural Failure and Fire Passage	84
Potential Fuel Source	84
Combustion Emissions	85
Design Implications	85

Wind	85
Moisture	86
Mold and Mildew	87
Interior Humidity	88
Damage to Wall or Interior	88
Design Implications	91
Insects	91
Subterranean Termites	91
Carpenter Ants	93
References	94

Chapter 5. System Selection — 97

Important Criteria Not Quantified	97
Availability	100
Service	100
R-values	100
Dimension Consistency	100
Price	101
Concrete Content	101
Unit Size	102
Foam Webbing	102
Fastening Surfaces	103
Foam Class	104
Specialty Units	105
Unit Assembly	106
Research Reports	106

Chapter 6. Building Envelope — 109

Important Note on Format	109
Dimensioning	109
Wall Thickness	111
Wall Alignment	115
ICF Levels	116
All-ICF	117
All ICF except Roof Ends	119
ICF Basement Only	124
Stem Wall Only	125
All Above-Grade Walls	125
ICF First Story Only	128
Insect Provisions	128
Subterranean Termites	128
Carpenter Ants	134
General Steps	134
Spans and Projections	134
Divison of Openings	135
Window and Door Mounting	139
Window Details	146

Chapter 7. Attachments — 153

- Floor Decks — 153
  - Material — 153
  - Frame Decks — 154
  - Concrete Floors — 159
- Roofs — 162
- Interior Walls — 164
- Utility Lines — 166
- Interior Finish — 171
- Exterior Finish — 171
- References — 176

Chapter 8. Cost Estimation — 177

- Sensitivity Estimation — 177
  - Lumber Price — 178
  - Concrete Price — 178
  - Formwork Price — 178
  - Exterior Finish — 178
  - Curved Walls — 180
  - Spans and Projections — 180
  - Disaster Provisions — 180
  - HVAC Sizing — 181
  - Air Change — 181
- Itemized Estimation — 181
  - Step One: Itemization — 186
  - Step Two: Estimating Waste Factors — 192
  - Step Three: Estimating Prices — 193
  - Step Four: Calculating Total Item Costs — 193
  - Step Five: Calculating Total Project Costs — 193

Part 4 Engineering — 195

Chapter 9. Structural Performance — 197

- Design Extremes — 197
- Wind — 199
- Earthquake — 200
- References — 201

Chapter 10. Structural Design — 203

- Overview — 203
- Structural Checks — 204
- Reinforced Concrete Design Procedures — 206
  - Minimum Reinforcement — 207
  - Flexure and Axial Loads — 207
  - Empirical Design Method — 207
  - Walls Designed as Compression Members — 208
  - Slenderness — 210

Perpendicular Shear in Walls	214
Parallel Shear in Walls	217
Lintel Bending	217
Lintel Shear	219
Reinforced Concrete Example	222
Shear Perpendicular to Wall	222
Shear Parallel to Wall	222
Axial and Flexural Loading	225
Structural Plain Concrete Design Procedures	231
Perpendicular and Parallel Shear in Walls	232
Axial and Flexural Loading	232
Applicability	233
Structural Plain Concrete Example	235

Chapter 11. Energy Efficiency and HVAC 241

R-value	241
Air Infiltration	242
Thermal Mass	243
Combined Effects	245
Correction for Thermal Mass Effect	245
Correction for Air Infiltration	247
Code Treatment	248
Sizing Equipment	249
Empirical Method	249
Adjusted Calculation	249
Ventilation Provisions	250
References	256

Chapter 12. Possible Future Developments 253

Concrete Compressive Strength	253
Minimum Reinforcement	254
Ongoing Research	255
Use of Structural Plain Concrete Provisions	255
Advanced HVAC Calculation	256
References	256

Part 5 Assembly 259

Chapter 13. Process Overview 261

Chapter 14. Formwork 271

Crew Composition	271
Dowel Placement	272
Unit Orientation	273
Course Leveling	275
Set Sequence	277

Bond Pattern	279
Interconnection	281
Bracing	283
Scaffolding Alternatives	284
Curved Walls	284
Irregular Angles	288
Opening Formation	288
Irregular Bucks	291

Chapter 15. Concrete 293

Mix	293
Placement Equipment	295
Consolidation	299
Sloped Pours	300

Chapter 16. Attachments 305

Floor Joist Pocketing	305
Surface Cutting	308

Appendix A. Metric Conversion Factors   311

Appendix B. Canadian Equivalents of Referenced U.S. Standards   315

Appendix C. Directory of Product and Information Sources   317

Index   321

Introduction

This book is for professionals interested in designing and building houses out of insulating concrete forms (ICFs). ICFs are hollow units made of plastic foam that are assembled into the shape of a building's exterior walls. Much like conventional formwork, ICFs are filled with reinforced concrete to create structural walls. However, they are left in place to provide insulation and a surface for finishes. Figures I-1 through I-9 show typical steps in the construction process.

Compared with wood frame, ICF walls have advantages in energy efficiency, interior comfort, sound attenuation, design flexibility, strength, durability, disaster resistance, pest resistance, price stability, and ease of construction. These account for the systems' rapid growth in North America over the past 5 years. The cost of construction with ICFs is currently slightly higher than that of frame, but it is declining with design improvements and builder experience.

Figure I-1 Setting forms on the footing. (*R.P. Watkins Inc.*)

Figure I-2 Completed first course of formwork. (*R.P. Watkins Inc.*)

Figure I-3 Completed basement formwork. (*American Polysteel Forms.*)

Most ICF buildings to date have been single-family houses, and this book is limited to considering that application. However commercial, institutional, and industrial buildings have also been constructed of the systems, and many of the data and principles presented here should be useful for these as well.

The book is written, above all, to meet the needs of architects and engineers designing ICF houses. Its focus is design principles and formulas, and design and performance data. It also describes field assembly so that designers can better understand how ICFs go together in the field and can specify aspects of installation that they

Introduction   xiii

Figure I-4 Pouring concrete into the basement formwork. (*Lite-Form Inc.*)

Figure I-5 Cutting formwork for above-grade walls.

Figure I-6 Setting formwork for above-grade walls.

Figure I-7 Framing interior walls and roof.

Introduction xv

Figure I-8 Preparing completed ICF walls for finishes.

Figure I-9 Completed house.

consider critical. The book may also be of interest to contractors who want to know the design and engineering details of the systems.

The book focuses on U.S. products and building practices. However, extra information relevant to professionals operating in Canada is included. Most discussion and data include only brands of ICFs currently distributed in the United States. Several of these are also popular in Canada, but some other brands are sold exclusively north of the border. For Canadian readers we list these additional brands in Appendix C at the back of the book. Most home construction practice in the two countries is highly similar, and much of the content of the book is transferable. However local convention, the more northern climate, and a few sharp code differences have led to some home construction preferences and requirements that diverge from those in the United States. We note some of the major differences in the text. Select tables include data in metric units, and Appendix A contains metric conversion factors. Appendix B lists the nearest Canadian equivalents for U.S. codes and standards referenced in the book. Note, however, that it is always the reader's responsibility to determine what information does not apply to his or her circumstances and act accordingly.

The text avoids repeating material already available in other publications. Most importantly, it is superseded by the product documentation of the ICF manufacturers. The book's design and assembly guidelines are intended only to provide general principles of ICFs that transcend brand, or to fill in details not addressed by manufacturers' information. The rules and recommendations of the manufacturers always take precedence over this book.

In addition, the material included in *The Insulating Concrete Forms Construction Manual* (McGraw-Hill 1995) is not much discussed here. That manual provides step-by-step instructions on field assembly for homes built of ICFs. To keep the directions simple it presents the most-used or most-recommended building practices only, and goes into detail on how to perform these in the field. The assembly section of this book concentrates instead on presenting and evaluating the important alternative practices for each major step of construction so that the designer and contractor can choose among them.

This book is based on the reported practices of actual design and construction professionals and the published technical information of product manufacturers and government, industry, and regulatory organizations. It is intended for the use of building professionals who are competent to evaluate the significance and limitations of this information and who will accept responsibility for use of the material it contains. The Portland Cement Association is not responsible for application of the material contained in the book or for the accuracy of the material provided by any of the information sources.

Chapter 1 ("ICF Characteristics") of Part 1 ("Performance") describes the attributes and behavior of ICF structures to date, as measured in the laboratory and observed in the field. ICFs perform in ways that differ from (and frequently exceed) simple extrapolations of traditional formulas and design rules. Examples are their structural capabilities and energy efficiency. In Chapter 1 we discuss the factors that account for these differences and direct the reader to the later sections of the book that treat them in detail.

Part 2 ("Components") breaks the available ICF systems into their parts and materials. Chapter 2 ("ICF Systems") provides the terminology used to identify and describe each brand and its pieces. Chapter 3 ("Plastic Foams") presents data on the predominant material of the ICFs, plastic foams.

Part 3 ("Design") presents the information necessary to the tasks normally performed by the architect. Chapter 4 ("Detailing Considerations") covers how important phenomena (fire, wind, and so on) can impact ICF houses and what steps are taken in design to account for them. Chapter 5 ("System Selection") presents criteria and data for choosing the best ICF for a specific project. In Chapter 6 ("Building Envelope") are alternatives and recommendations for doing layout and specifying wall design; and in Chapter 7 ("Attachments") for specifying structural connections and aesthetic and finish details. Chapter 8 ("Cost Estimation") covers methods of estimating total and itemized job costs.

Part 4 ("Engineering") contains rules, formulas, and data for structural and HVAC engineers. Chapter 9 ("Structural Performance") summarizes the structural capabilities and performance of ICFs based on their record in previously built houses. Procedures for performing structural engineering on houses under design are in Chapter 10 ("Structural Design"). Chapter 11 ("Energy Efficiency and HVAC") gives guidance for calculating the load requirements of the HVAC system. Ongoing testing and research are suggesting refinements in ICF structural and HVAC engineering; Chapter 12 ("Possible Future Developments") discusses the most likely near-term changes.

Part 5 ("Assembly") covers the tasks performed by the contractors. Chapter 13 ("Process Overview") is a brief description of the typical sequence of construction. Chapters 14 ("Formwork"), 15 ("Concrete"), and 16 ("Attachments") cover the alternative methods of assembling the foam formwork, selecting and placing concrete into it, and connecting other key members and components to the walls.

As with any construction, the three major tasks represented by

Parts 3 through 5 affect one another. Design decisions have logistic impacts on assembly. Engineering calculations sometimes reveal the need to change the planned layout. The text of these chapters highlights such interactions so that decisions at one stage of the process can include consideration of their impact on the others.

As noted earlier, Appendix A contains metric conversion factors, and Appendix B the identity of relevant Canadian codes and standards. Appendix C provides contact information on key sources of products and technical assistance, including the ICF manufacturers and relevant trade associations.

An 8-page section of color photographs at the center of the book depicts the wide range of types and styles of homes constructed with ICFs to date.

Like any new construction system, the ICFs are undergoing rapid change as manufacturers improve their designs and installers refine their technique. Fortunately, the available body of design and test data has grown rapidly in recent years. We attempt to present here the most recent possible data and information. However, designers must always check with the manufacturer of the system being used and follow its recommendations. Manufacturers' recommendations always take precedence over any information in this book.

Field experience and lab data suggest that ICFs have the potential to provide a superior home at a competitive price. The authors hope that this book helps building professionals realize that potential.

Acknowledgments

Production of this book involved the cooperation of hundreds of people.

Dan Mistick of the Portland Cement Association originated the project, organized it, and kept it on track throughout much of its lifetime. He, Dick Schmickle, and George Barney also found the resources for it and forged the alliance of corporations and trade associations that provided the masses of information on which it draws. James Farney gave extensive review for style as well as content.

The Canadian Portland Cement Association gave extensive support. Norm MacLeod polled Canadian building professionals about what they would like to see in such a book. He also put us in contact with many other helpful information sources. Cameron Ridsdale was a tireless and merciless reviewer who also happens to be a talented architect and an expert in this field. He dug up important new information on several topics, and is the principal author of the section on concrete floors. P. L. Maillard meticulously checked technical sections. Alicje Cornelissen did all of the metric conversions. Kathleen Gissing helped track down several Canadian ICF manufacturers.

Dick Whittaker, President of the Insulating Concrete Form Association, provided wise advice concerning the overall direction of the project as well as insightful editing and liaison with the manufacturers. Mike Eckert of the National Ready Mixed Concrete Association reviewed the complete draft and made important corrections from his industry's perspective.

Every one of the manufacturers of ICF systems covered in the book was extremely helpful, supplying materials and reading sections for accuracy. We were highly impressed by the extent of their cooperation with an independent project. Perhaps they were not afraid of a book designed to tell the whole story of their products because they have such a good story to tell.

A half-dozen manufacturers of plastic foams devoted technical and marketing staff to advise us, provide documents, and review drafts.

Thanks particularly to Alisa Hoffee and Jack Lubker of Amoco, Rebecca Liebert of ARCO, Tom Greeley and Gene Zimmerman of BASF, Jim Shannon of Huntsman Chemical, and Susan Herrenbruck, Lisa McKellar, and Fran Lichtenberg of the Society for the Plastics Industry.

There were more contributors from architectural and engineering firms, from contracting and concrete companies, from research laboratories and universities and code and testing organizations than we can count, let alone name. However it seems fair to single out Comfort Homes of Lusby, MD, and EPS Building Systems of Harrisburg, OR. Their people spent days with us covering particulars of design and construction and providing detailed field data.

At the publisher, thanks must go to Zoe G. Foundotos and Peggy Lamb. Peggy is McGraw-Hill's longtime editor extraordinaire. Her high standards and meticulous attention to detail brought clarity, consistency, and class to both the prose and the visual presentation.

Our remaining thank-yous go to Shari VanderWerf. She organized, edited, kept track, reminded, located, advised, verified, tracked down, reread, found, proofed, and checked. She is also the best wife any author could hope to have.

The authors are entirely responsible for any omissions or errors in this book. The many other contributors are largely responsible for its inclusions and its usefulness.

Part 1

Performance

The attraction of insulating concrete forms (ICFs) is that they outperform other wall systems in many desirable ways. Although currently their initial cost is slightly above that of conventional wood frame, a growing number of building professionals and home buyers have decided that the benefits and life-cycle savings of ICFs outweigh that premium. Understanding the performance differences precisely is important to deciding when and where to use ICFs. It is also critical background for designing with them effectively.

Chapter 1

ICF Characteristics

We attempted to gather precise current information on the performance of houses built of ICFs. Summarized in this chapter, that information should help in deciding whether to use ICFs in particular projects. The chapter also mentions the important design implications of the systems' unusual characteristics. Many of the topics treated here are covered in more depth in later sections of the book. Appropriate references identify these for the interested reader.

Information about ICF performance is increasing and changing as experience accumulates. The stock of ICF houses is growing, designers are pushing the boundaries of their uses of the systems, and manufacturers are advancing their product designs. We therefore point out important expected future developments, in addition to describing what is known as of the spring of 1996.

Cost

Current construction cost of ICF walls is slightly above that of frame. However, initial savings that result in other areas of construction [notably heating, ventilation, and air conditioning (HVAC)] sometimes offset much of this premium; the premium itself is declining over time, and important operating costs of an ICF home are lower.

Wall construction cost

The construction cost of homes built with ICF exterior walls by experienced crews is currently approximately $1.00 to $4.00 per square foot of gross exterior wall area above the cost of building the same structure with conventional wood frame. Note that this is a comparison to fully ordinary frame. No adjustment is made here to include the extra costs of modifying the frame wall so that it has the benefits of ICF walls, and, in fact, that might be impossible.

4   Performance

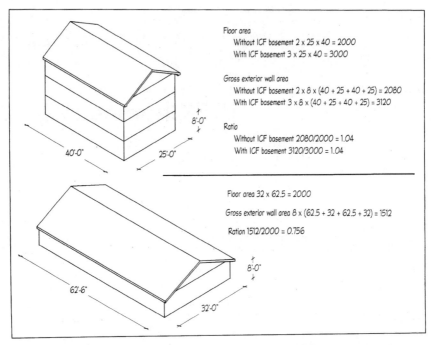

Figure 1-1   Relationships between gross exterior wall area and floor area.

For those more accustomed to working with floor space than with gross exterior wall area, there is a rough, handy conversion. We depict it in Figure 1-1. In most multistory houses, the ratio of enclosed floor area to gross exterior wall area is approximately 1:1. Therefore the extra cost per square foot of floor area is still estimated at about $1.00 to $4.00. In the average one-story house (approximately 2000 square feet of floor space) the floor-to-wall ratio is closer to 1.33:1. Therefore the incremental cost per square foot of floor space will be approximately $0.75 to $3.00.

The typical breakdown of ICF wall costs presented in Table 1-1 gives an appreciation of the major cost components. These data are in the middle of quotes we received from 20 contractors on actual projects they completed in 1995 or 1996. Actual costs of any specific project will depend on many factors, notably local labor and materials prices, site logistics, and specifics of design. Chapter 8 details the procedure for preparing more detailed cost estimates for specific projects.

Important influences on wall cost

Separate from regional variations in the costs of labor and materials, a few identifiable factors have a large influence on whether the costs of a

ICF Characteristics

TABLE 1-1 Representative Costs of ICF Wall Construction

Item	Cost per gross square foot of wall area*
ICF units and accessories	$2.50
Concrete	0.80
Rebar	0.30
Lumber	0.40
Other materials	0.10
Labor	1.25
Total	$5.35

*Costs vary widely with local labor and materials prices, particulars of building design, and other factors. These figures are near the middle of the range reported by U.S. builders during spring 1996 from actual recent projects.

TABLE 1-2 Important Factors Affecting Relative ICF Wall Construction Cost

Decreases Cost of ICF Homes Compared with Frame
Low concrete prices (versus typical national figure of $50/cu yd)
Low ICF form prices* (versus typical 1996 figure of $2.50/sq ft)
High lumber prices (versus 1996 figure of $300/1000 board feet)
Design specifies stucco as exterior finish
Design must provide for high-wind conditions
Design includes curved walls

Increases Cost of ICF Homes Compared with Frame
High concrete prices
High ICF form prices
Low lumber prices
Design specifies shingles or shakes as exterior finish
Design includes large wall projections or wall spans
Design provides for high earthquake hazard

*These estimates do not include potential cost savings from added features of more expensive systems.

particular ICF project will be nearer the low or the high end of the range cited here ($1 to $4 above frame). These are listed in Table 1-2. A project with many of the attributes that lower the relative cost of ICF construction (as noted in the table) and few of the attributes that raise it will tend to be near the bottom end of the cost range, and vice versa.

Note that some of the factors listed do not affect the cost of the ICF wall itself, but of other components. For example, the use of stucco reduces the relative cost of ICFs because stucco can be placed onto them less expensively than onto frame. (Likewise, the high energy efficiency of ICFs allows the HVAC equipment to be smaller, and therefore less expensive. See "Construction Cost Savings.")

Note also that all of these estimates assume the use of an experienced ICF wall crew. Crews are reportedly "experienced" after constructing three or four houses. If the crew has done fewer, the cost premium will, on average, be higher than it would be otherwise because of the slower work rate of the inexperienced workers and some greater waste.

Chapter 8 includes a more detailed discussion of major influences on the total cost, and of methods that take the many factors into account to produce more precise estimates for specific projects.

Construction cost savings

Using ICFs can permit initial cost savings in parts of the structure other than the exterior walls. In some houses these various savings are believed to have offset most of the extra cost of building with ICFs. The greatest of these is in the HVAC equipment. Because of the greater energy efficiency of ICF walls, engineers have downsized the HVAC capacity by as much as 50 percent compared with what they would have installed in frame houses. However, since many HVAC contractors unfamiliar with ICFs tend to size components as they would in a conventional house, the savings depend on the use of a contractor capable of correctly factoring in superinsulation. Chapter 11 discusses how sizing is performed.

Some builders claim that the greater bearing capacity of ICF exterior walls allows them to use fewer load-bearing interior walls. However, there is no quantification of the savings from this.

Cost trends

In the past ICF construction costs have been falling for two major reasons. First, the manufacturers have improved their designs in ways that reduce installation labor. Examples are special ICF units that eliminate the need to make cuts for occasional wall details (such as corners and brick ledges) and special surfaces on the ICF units that are capable of holding conventional wood fasteners. Second, field crews are inventing new, more efficient assembly procedures that spread throughout the industry. This goes beyond the savings of improved efficiency resulting from practice. Many of the design details of Chapters 6 and 7 and assembly options in Chapters 14 to 16 originated in the field.

In the future, costs should continue to fall for both of these reasons, plus one more. As the production volume of ICF units increases, manufacturers should realize economies of scale. According to figures from foam molding companies, in very large volume the ICFs might drop in cost by as much as 40 percent (about $1.00 per square foot). Although firm projections are impossible, the Portland Cement Association (1995) has forecast that if current growth rates in ICFs continue, by the year 2005 their total installed cost could be virtually the same as that of frame construction.

Operating costs

The superior attributes of ICF walls can result in lower operating costs for their owners. The greatest savings usually come from a reduction in heating and cooling costs and lower insurance premiums. Reported HVAC bills from ICF homes have averaged roughly 25 to 50 percent less than those of comparable frame houses nearby. Most property insurance providers offer a premium reduction for "superior construction" or high fire or wind resistance, for which concrete walls typically qualify. The savings on an average home can be $40 to $100 per year.

Energy Efficiency

Comparison with similarly sized frame houses next door suggests that constructing the exterior walls of a home from ICFs reduces HVAC energy consumption by 25 to 50 percent, and in some cases more. There are three distinct factors contributing to the overall reduction: R-value, air infiltration, and thermal mass. Table 1-3 compares approximate values for these across wall systems.

R-value

The R-value measures the tendency of a material or assembly (such as a wall) to slow the passage of conducted heat, as depicted in Figure 1-2. Conduction through the exterior envelope is usually the single largest source of unwanted heat loss or gain in modern homes, and conduction through the walls is a major portion of the total. The R-value of an ICF wall runs more than double that of a 2×4 frame wall, and about 50 percent more than a 2×6 frame. In other words, heat conducts through the walls of a typical ICF about half as fast as through a typical 2×4 wall and two-thirds as fast as through a 2×6.

TABLE 1-3 Approximate Energy Performance Data for Alternative Wall Systems

Property	Measure	Typical value for wall of:		
		ICF	2×4	2×6
Resistance to conduction	R	18–35	10–12	15–19
Air infiltration	ACH (for house)	0.10–0.35	0.3–0.7*	0.3–0.7*
Thermal mass	Btu/(sq ft · °F)	5–20	2–3	3–4

*Air infiltration data usually do not discriminate between 2×4 and 2×6 walls.

8 Performance

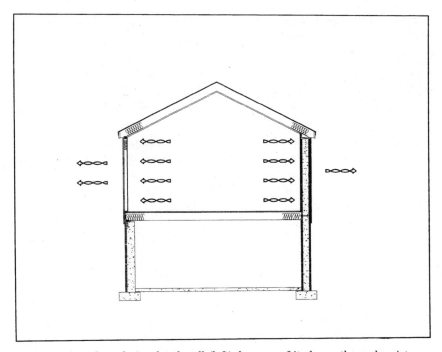

Figure 1-2 A moderately insulated wall (left), because of its lower thermal resistance, allows the outflow of more heat than a highly insulated wall (right).

This is as measured in the guarded hot box test, the method generally regarded as most thorough and precise. Various "calculated" R-values are frequently also reported for all types of systems. However, comparing these can be hazardous since they do not produce identical (or even always close) results, and are therefore open to selective reporting. Chapter 11 covers this issue in more detail.

Air infiltration

Outdoor air that enters through gaps in the envelope must be heated or cooled to indoor temperature, as depicted in Figure 1-3. This is a second major contributor to the heating and cooling load of houses (ASHRAE, 1993).

A standard measure of the rate of air infiltration into a home is air changes per hour (ACH), usually determined with the so-called blower door test. Tests of houses with mostly or all ICF exterior walls found ACHs at ambient air pressures of 0.12 to 0.35. This compares with measures of approximately 0.3 to 0.7 ACH for new frame houses, as documented in various surveys (ASHRAE, 1993). For both wall systems, the lower air infiltrations were apparently achieved by the

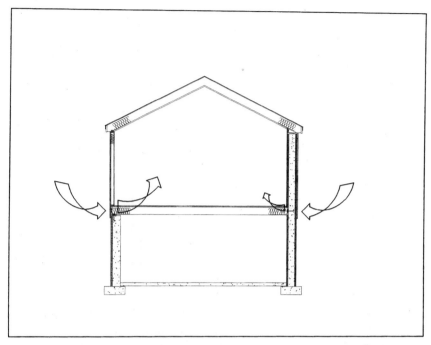

Figure 1-3 Higher air infiltration of a moderately tight wall (left) requires more energy to condition the incoming air than the lower air infiltration of a very tight wall (right).

use of special additional measures taken during construction to increase the tightness of the house.

The tightness of ICF houses also recalls the need in modern construction to consider proper ventilation for maintenance of indoor air quality. There are straightforward modifications to the HVAC system to accomplish this, which are covered in Chapter 11.

Thermal mass

Thermal mass is an informal term for the ability of materials to absorb and store heat without great increases in their own temperatures. For walls it can be measured as the heat capacity per square foot of wall area, or Btu/(sq ft · °F). Walls with great thermal mass tend to buffer the interior of a home from the extremes of outdoor temperature that occur over the cycle of day and night. Figure 1-4 depicts this effect. It acts to reduce both peak and total HVAC loads.

As quantified in Table 1-3, the median ICF wall has several times the thermal mass of a standard frame wall. This results in energy savings over and above those resulting from the ICF wall's higher R-value and lower air infiltration. As one might expect, the effect and

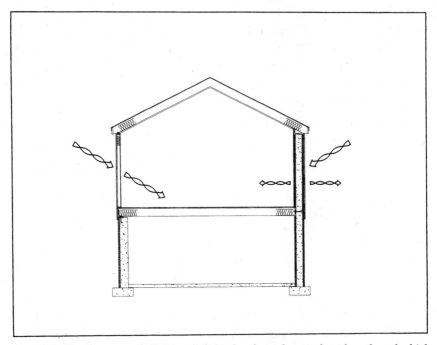

Figure 1-4 The low mass of a light wall (left) absorbs and stores heat less than the high mass of a heavy wall (right).

the savings are greater in warm climates, where the temperature fluctuates around the internal HVAC set point for a large part of the year. Quantifying the savings is complex. But we can give an intuitive explanation of them.

Engineering calculations suggest that the savings of replacing standard 2×4 exterior frame walls with ICF walls is, in a temperate climate, approximately equal to the savings of replacing that standard wall with an R-50 frame wall. In very cold climates the savings from thermal mass are lower, closer to the savings of going to an R-30 frame wall. Chapter 11 provides more detail on the effect and the mechanics of its measurement.

Combined effect

The combination of higher R-value, lower air infiltration, and higher thermal mass is believed to account for the 25 to 50 percent energy savings of ICF versus frame homes. Informal inspection of the field data suggests that these factors also account for the variation in savings: the greatest savings tend to occur in houses using the highest-R

systems (R-value effect) located in the most temperate climates (thermal mass). Some attempts are under way to combine some or all of these effects into a single measure of energy efficiency, as described in Chapter 11.

Code and regulatory treatment

ICF walls meet the R-value requirements of virtually every residential building code in the United States without modification. In addition, a growing number of codes give a form of credit to them for their thermal mass. High-mass walls, in effect, are assigned higher R-values for code purposes than traditional measures and tests allow, to account for the energy savings of a building with high-mass walls. This is not usually important since the walls (as noted) already meet the code. However, if other parts of the envelope are below required R-values, some codes allow compensating with the parts that exceed the code. In these cases the credit for thermal mass may prove helpful.

The walls are also effective in meeting the requirements of various energy-efficiency incentive programs. The Canadian government sponsors the R2000 program, which maintains a set of particularly high standards for home energy efficiency, air quality, and other important features. Homes receiving an R2000 certification go through design review, several inspections, and physical testing. Canadian builders report that their frame walls require substantial additional detail work (such as extra insulation, special infiltration barriers, caulking) to meet the standards, whereas their ICF walls require nothing extra. U.S. builders report the same when they attempt to meet the standards of various utility-sponsored programs.

Design Versatility

ICF homes are now being constructed with virtually every type of design feature popular in frame. The color photographs in the center of the book demonstrate this with examples of both architecturally novel houses and homes that blend into conventional neighborhoods. This design versatility contrasts with older concrete construction technologies, which, by the force of costs, tended to constrain design to rectangular footprints, rectangular openings, flat surfaces, and perfectly vertical walls (no upper-story projections or setbacks). Some attractive irregular features actually add less cost to ICF construction than to frame. This is because workers form the shape of ICF walls by cutting and assembling a soft, easily manipulated material (foam).

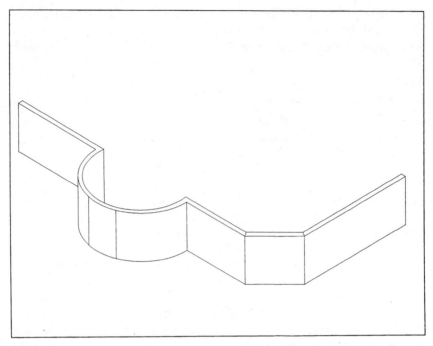

Figure 1-5 House with footprint irregularities: curved walls, non-90° corners, frequent corners.

Footprint

Generally speaking, footprint variations are as easy or easier with ICFs than with frame. Figure 1-5 depicts the variations of most interest. The incremental cost of a corner on an ICF house is, by all estimates, virtually equal to the incremental cost of corners on frame. As with frame, a 90° angle is formed simply by abutting the appropriate standard components. Irregular angles have historically followed the same cost rule. They are usually made by miter cutting the forms. However, new, continuously adjustable ICF corner units hold promise for making irregular angle formation faster and less expensive (see Chapter 2).

One attractive footprint variation is easier with ICFs: curves. As described in Chapter 14, multiple methods exist for forming ICF units into an arc and stacking them into a curved wall.

Openings

Door and window openings are also comparable with either system, whether they are rectangular or irregular. Rectangular ICF openings are usually formed as stacking proceeds, much as with concrete mason-

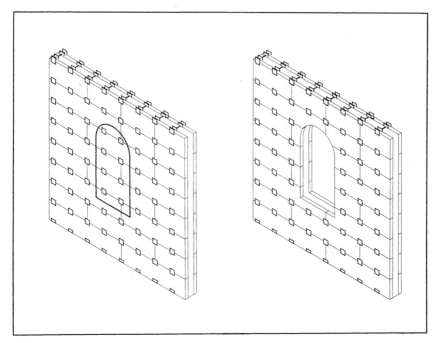

Figure 1-6 Cutting an irregular opening out of ICF formwork in place.

ry. Irregular openings of almost any shape are quickly and easily cut out of a stacked wall of ICF formwork in place, as illustrated in Figure 1-6.

Surface relief

Relief on the wall surface can be created as or more easily than on frame. Wooden or rigid plastic trim is fastened in the same way. When stucco is the finish material, relief on both systems is generally accomplished by adhering shaped pieces of foam to the wall surface (Figure 1-7), then finishing over. The only significant difference is that frame requires a sheathing (usually a thin layer of foam) to provide a substrate for stucco. ICFs do not because they already have a foam surface.

Projections and spans

Although past designers of ICF houses avoided the sort of projections and spans depicted in Figure 1-8, recent houses have exhibited a broad range of upper-story garrisons, bays, and setbacks, which provide interesting variations in floor plan and wall depth. Some examples are among the color photographs in the center of the book. Such features still usually add more cost to ICF construction than to frame,

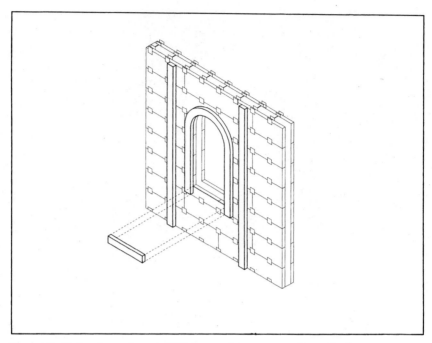

Figure 1-7 Gluing foam pieces to ICF formwork to create surface relief.

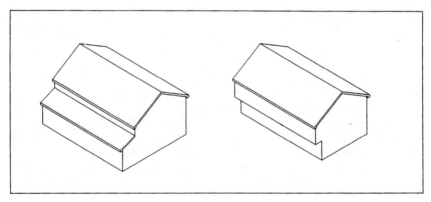

Figure 1-8 Examples of upper-story spans (left) and projections (right) not supported below.

but the premium appears to be declining and already acceptable in many projects.

The earlier reluctance and higher cost stemmed from the logistical and structural challenges presented by upper-story walls that are unsupported by a wall below. Because of the weight of the concrete, it is not feasible to support ICF walls on a conventional lumber floor, as is done with frame. However, solutions have appeared. For one, it is possible to construct the floating portions of the upper-story wall from frame. Details are given in Chapter 6. Most recently it has also proven possible to leave the wall all ICF, but support it through greater use of reinforcement or a concrete or steel floor. Chapters 6 and 9 discuss this option. The use of upper-story ICF walls retains the benefits of the foam and concrete throughout the exterior walls.

Construction Logistics

A striking advantage of ICFs over other innovative new wall systems is the ease with which traditional contractors learn it and the speed with which they come to like it. The vast majority of the contractors building ICF walls are carpenters who also build with frame. They use most of the same tools they always have, although sometimes in slightly new ways. Even though the tasks of erecting the formwork are not highly similar to those of framing, they are intuitive and easily learned. The work is light, and some complex shapes are more easily formed than with frame. Thus the workers typically tire less. Ordering and placing concrete is familiar to many of these crews from such past work as building footings, although they will typically need some help in this area at first. All of these tasks are discussed in Part 5. As mentioned previously, most crews appear to master the basics after three or four houses.

The other trades have little adjustment to make. The electrician must alter standard practice the most. Creating channels and cutouts for electrical boxes requires cutting foam or running conduit, instead of boring lumber. However, the effort and cost involved are, by all accounts, indistinguishable from work on frame. The few electrical contractors who initially want to charge extra for ICF work can generally be talked out of it after they have been shown the simple mechanics. These are described in Chapter 16.

Structural Capabilities

Like conventional reinforced concrete, ICF walls are readily designed with strengths far beyond the requirements of most low-rise construction. One can economically adapt them to conditions of high load factors, multiple stories, and wide openings by varying specifications on the concrete and reinforcing steel. Chapter 9 discusses in more detail the forms achieved and loads endured to date.

Performance under extreme loads is expected in many situations to exceed that of frame, based on past experience with similar reinforced structures. The only structural forms common to frame that have not routinely been handled with ICFs are the projections and spans discussed earlier in this chapter. However, there are now multiple methods of creating them, and the cost and original engineering work necessary to doing so are declining as experience accumulates.

Importantly, recent calculations and engineering data suggest that much ICF engineering practice is especially conservative. Current procedures are adaptations of the conventional reinforced concrete guidelines of the American Concrete Institute (1996) and the Canadian Standards Association. But these were written primarily for larger structures built with formwork that does not cure concrete in the near-optimal fashion that foam does. Testing and new analyses designed to determine the structural capacities of ICF walls more accurately and to explore whether simpler and less expensive construction techniques would be adequate are under way at Construction Technology Laboratories and the Portland Cement Association.

In Chapter 10 we present details of currently accepted structural engineering practices. Chapter 12 describes the outstanding issues, the research into them, and the possible changes in future practice that may result.

Disaster Resistance

The evidence indicates that ICFs are relatively resistant to the ill effects of natural disasters. The most costly types of disasters, measured in homes affected, injury and loss of life, or dollars of property damage, are fires, high winds, and earthquakes, in that order.

Fire

In fire wall (ASTM, 1995a) tests, ICF walls withstood exposure to intense flame without structural failure significantly longer than do common frame walls. They also prevented the passage of fire for a longer period of time. These measurements are consistent with the fire resistance of reinforced concrete structures observed in the past.

The foams used in most ICF forms are treated so that they will not support combustion, unlike lumber. In the event that they are held in the presence of an outside flame source, their tendency to transmit flame as measured by the flame spread index (ASTM, 1995b) tests at less than half that of most woods. Some of the foams can produce more visible smoke than lumber when burned (as measured in tests by their smoke developed index), although they are still well within limits allowed by standards and building code organizations. Chapter 4 presents the fire performance data and recommended design practice in more detail.

Wind

Surveys of major hurricane sites (NAHB Research Center, 1993; Zollo, 1994) reveal a discernibly higher structural survival rate for reinforced concrete walls subjected to high winds than for conventional frame. This high wind resistance results from inherent properties of the wall system: the great weight of the concrete and the high, continuous tensile strength of the steel reinforcement that extends from footing to roofline.

Roof failures in wind are actually more frequent than wall failures, and one might assume that the use of ICF walls could not mitigate this. However, certain simple, inexpensive details for the ICF-roof connection can add a measure of resistance. Specifically, use of steel strapping embedded in the concrete to attach the roof members replaces a great weak point (the nailed lumber-to-lumber connection) of the assembly with a much stronger one. Chapter 4 provides more depth on wind damage, and Chapter 7 gives the related design details.

Earthquake

Earthquake is the only major natural disaster to which concrete is not inherently more resistant than conventional frame. Particularly the unreinforced and lightly reinforced low-rise concrete buildings of the distant past are believed to have suffered more damage in earthquakes. New, more conservative reinforcement schedules are now recommended and applied in areas of high seismic risk, and are expected to perform well. Indeed, some ICF homes have survived recent southwestern quakes without significant damage.

Chapter 9 expands on the available information regarding the performance of wall systems in earthquakes.

Moisture Movement

ICF basements are subject to the same groundwater forces as conventional basement walls, and merit similar treatment to resist water penetration.

Above-grade moisture problems have been generally absent from ICF structures, and there are logical reasons why many moisture phenomena that have plagued some frame buildings will be rarer with ICF walls. Nonetheless, there are good reasons for practicing conservative design in certain situations.

The continuous insulation reduces wall cold spots and most of the potential for interior condensation. Such condensation is blamed for interior finish damage and mold and mildew in some wood and (especially) steel frame buildings.

As with frame construction, the exterior finish of an ICF wall is supposed to prevent water from contacting the structure and insulation. However, if the finish fails, the concrete and foam are resistant to damage. In contrast, wood is prone to rotting. Water entering the concrete and foam envelope also has few likely paths for reaching the interior (Bahr, 1995; Kenney, 1995). Experimental results suggest that it is more likely to drain down than inward. But given the newness of the system, it is advisable to take as much care with exterior finish and moisture provisions as one takes with frame. The details of Chapter 6 include recommended provisions.

We have seen documentation of moisture penetration through ICF walls from one source. In rare instances, rainwater driven against windows set into walls finished with stucco has reached the interior of a home. Similar problems have previously appeared in frame houses, where they have caused consequential damage (Energy Design Update, 1995). Analysis suggests that failures of the seal between window and wall allow water under pressure to move inward between the two assemblies. The window details of Chapter 6 include provisions intended to reduce the risk of such a failure.

Chapter 4 includes more complete discussion of the moisture resistance and critical connection points of ICF walls.

Insect Resistance

To date we have been unable to document a single instance of insect infestation of an ICF wall. There are also reasons to believe it would be unlikely, particularly in a house with all-ICF exterior walls. However, there have been instances of insect penetration into the foams of other wall systems (Andrews, 1992; Mar-Quest Research, 1993; Smith and Zungoli, 1995). For this reason certain standard precautions are available for buildings located in high-risk areas.

As discussed in more detail in Chapter 4, the greatest concerns are that (1) subterranean termites will use the foam as a pathway to reach the wooden parts of a house, or (2) carpenter ants will nest in the foam as they do in wood. In neither instance is there an apparent risk to the concrete structure of the exterior walls, as there is with wood frame. The concern is that the foam will not stop the insect incursion, or in some cases could make it easier.

Figure 1-9 illustrates the common procedure for preventing subterranean termite invasion in a conventional house. An insecticide in the soil around the foundation blocks the insect's path. When the foundation is clad with foam, in theory the termite might burrow into it and up to the lumber above, avoiding the poison. Figure 1-10 depicts this.

ICF Characteristics 19

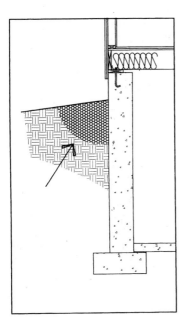

Figure 1-9 Path of subterranean termites blocked by insecticide around the foundation.

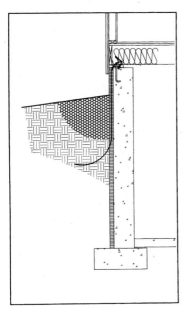

Figure 1-10 Theorized possible path of subterranean termites through a foundation's foam cladding.

It is not clear that a termite confronting an all-ICF house would detect the lumber of the roof, penetrate the dampproofing over the foam, and then burrow upward 8 to 20 feet to get its meal. However, for the cautious and those in high-risk situations, Chapters 4 and 6 present the mechanics of alternative preventive measures and appropriate details. As noted there, some codes now require one of these measures.

Carpenter ants have infested other foam-based wall systems. As discussed in Chapter 4, the same preventive steps used in frame apply to ICFs.

Durability

The long-term durability of any new type of wall assembly deserves scrutiny because full data are unavailable until the first buildings reach old age. Preliminary information on ICFs and analogous systems indicate good durability so long as proper precautions are taken with exterior finishes. The three areas to examine are the concrete itself, the foam, and the foam-concrete interface.

Concrete

Concrete foundations below grade have endured repeated and sustained exposure to groundwater, temperature cycles, and backfill and frost heave forces for decades without deterioration. Above grade, exposed concrete is similarly subjected to moisture, temperature, and freezing extremes with only slow deterioration and low maintenance.

Historically the aging mechanism assumed to cause the most damage to concrete is the freezing and thawing of absorbed moisture (Gonnerman, 1947). However, the frequency of the freeze-thaw cycles that concrete in an ICF will endure should be low because it is insulated.

Figure 1-11 demonstrates the point with the results of a thermal analysis of a typical ICF wall (Meilleur et al., 1995). Even at an outdoor temperature of $-11°F$, the concrete in the center is estimated to be only slightly below freezing. At higher outdoor temperatures the concrete should be above freezing.

Foam

Since the interior foam of ICF walls is covered with wallboard and rarely subject to extreme conditions or wear, its durability is not generally of concern.

We get some indication of the durability of plastic foams outdoors from a study by the Minnesota Department of Public Service (1988).

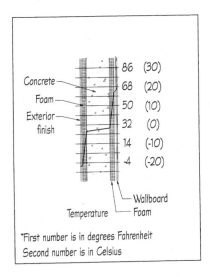

Figure 1-11 Temperature gradient across a finished ICF wall. (*Meilleur et al., 1995.*)

Field staff visited 59 separate Minnesota houses to note the physical condition of the foundation insulation, both below and above grade. The houses were 2 to 7 years old, and in all cases the insulation had been installed during construction. The foams were largely undamaged and undeteriorated. The major exception was gouging that had occurred either during construction (mostly below grade) or from lawn mower contact (above grade). In no case was such damage major. Above grade it always occurred on foam that had been covered with a relatively flimsy finish such as bitumen or paint alone. Moisture, ground and air freezing, and temperature cycles had no apparent effect. Moreover, depending on the exact type of foam, the insulations retained an average of over 90 percent of their rated R-values, despite taking on varying (but usually small) amounts of groundwater.

Experience with foam sheathings above grade has been similar. The foam layers of exterior insulation and finish systems (EIFSs) can experience physical damage in high-traffic areas if the stucco finish coatings are improperly or thinly applied. Then impact can cause gouges. Otherwise the foam is not generally known to deteriorate. The implication is that covering the foam properly is key. The details of Chapter 7 reflect this.

Foam-concrete interface

Field experience shows that the concrete poured into ICFs bonds with the foam. Depending on the variety of foam used, it can be extremely difficult to strip off. But one might be concerned that over time this connection will break as a result of differential expansion. This is not a structural problem. The foam is held in place by crosspieces and bears minimal weight for either interior or exterior finishes. However, a loss of connection could open a small gap for water and insects.

But such a break between foam and concrete appears unlikely. Thus far foams adhered to concrete over a decade ago still prove to be tightly bound. Moreover, the thermal expansion coefficients of both materials are extremely low—about 0.0000055 per degree Fahrenheit for concrete and 0.00003 for the foams most commonly used in ICFs. In addition, the temperature changes that the interface experiences should also be limited. As we can see in Figure 1-11, the temperature differential between materials is half that between indoor and outdoor temperatures. And the entire gradient from one face of foam through the concrete and into the next face of foam is small. The greater variation lies within the foam, a flexible material that is more likely to absorb differential expansions without damage.

Sound Attenuation

An attractive side benefit of ICF walls is their ability to restrict the passage of sound, as depicted in Figure 1-12. This reduces the intrusion of outdoor noise and enables ICFs to meet sound attenuation requirements of walls that separate living units.

The accepted measure of this property is the sound transmission class (STC). It is the output of a test that consists of generating sound over a range of frequencies on one side of a sample wall and measuring the fraction of the total sound energy that registers on the opposite side. Sound transmission is reported on a logarithmic scale. An STC of 10 corresponds to a little over one-third of the sound energy penetrating the wall, an STC of 20 to one-tenth, and so on. Table 1-4 relates these numbers to the occupant's perception of outdoor sound.

A conventional 2×4 wood frame wall filled with fiberglass and sided with wallboard on both sides measures STC = 36 (National Association of Home Builders, 1963). In contrast, various ICF walls have tested at 44 to 58. Thus compared to frame, most ICF walls allow less than one-third as much sound to pass through. The openings in a real wall (windows and doors) will reduce the STC for both wall systems, however, and bring the two complete envelopes somewhat closer together in practice.

Toxicity

Nothing held within or ordinarily emitted by an ICF wall is conventionally considered toxic. Depending on the precise contents, wet concrete

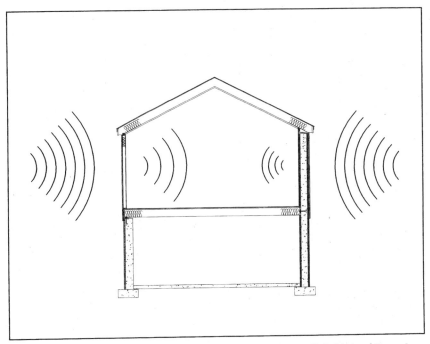

Figure 1-12 Sound passes more completely through a frame wall (left) than through an ICF wall (right).

TABLE 1-4 Effect of Wall STC on Sound Perception

STC	Audibility of loud speech on opposite side of wall
25	Easily understood
30	Fairly understandable
35	Audible but generally not intelligible
42	Audible as a murmur
45	Must strain to hear
48	Barely audible
50	Inaudible

SOURCE: Huntsman Chemical (1995).

can cause temporary skin irritation when contacted directly. Occupational Safety and Health Administration (OSHA) guidelines therefore require concrete workers to wear long clothing and gloves. Irritation from contact with hardened concrete is not reported, however.

None of the ICF foams appear to produce toxic reactions through contact, either. The evidence also indicates that they do not produce toxic emissions. The components of final ICF foams are considered benign; none include any formaldehyde. Moreover, the scientific measures of the air contents of actual houses insulated with the foams (discussed in more detail in Chapter 3) show an almost complete absence of any emissions.

References

American Concrete Institute, 1996
>   *ACI Manual of Concrete Practice, Part 3—1996*. Detroit, MI: American Concrete Institute.

Andrews, 1992
>   Andrews, Steve, *Foam-Core Panels and Building Systems*. Arlington, MA: Cutter Information Corp.

ASHRAE, 1993
>   *1993 ASHRAE Handbook of Fundamentals*. New York: American Society of Heating, Refrigeration, and Air Conditioning Engineers.

ASTM, 1995a
>   "Standard Test Method for Steady-State Thermal Performance of Building Assemblies by Means of a Guarded Hot Box," ASTM Standard C236-89. In *1995 Annual Book of ASTM Standards*. Philadelphia, PA: American Society for Testing and Materials.

ASTM, 1995b
>   "Standard Test Methods for Fire Tests of Building Construction and Materials," ASTM Standard E199-95a. In *1995 Annual Book of ASTM Standards*. Philadelphia, PA: American Society for Testing and Materials.

Bahr, 1995
>   Bahr, John, Construction Interface Services, Wilmington, NC. Personal communication.

Energy Design Update, 1995
>   "Severe Rotting Found in Homes with Exterior Insulation Systems," Vol. 15, No. 12 (December), pp. 1–3.

Gonnerman, 1947
>   Gonnerman, Harrison F., "Concrete." In Herbert F. Moore, *Textbook of the Materials of Engineering*, 7th ed. New York: McGraw-Hill.

Huntsman Chemical, 1995
>   Huntsman Chemical Corporation, EPS Technical Bulletin No. 4.30, Chesapeake, VA, August.

Kenney, 1995
>   Kenney, Russel J., R. J. Kenney Associates, Inc., Plainville, MA. Personal communication.

Mar-Quest Research, 1993
>   "Quantitative Analysis of Foundation Insulation's Impact on Subterranean Termite Control," Mar-Quest Research, Inc., Unpublished Report of Project No. 260-3, prepared for DowElanco, August.

Meilleur et al., 1995
>   Meilleur, Serge, André Grenon, Brad Savic, and Steve Allard, *Polycrete Installation Manual*. Longueuil, Quebec: Polycrete Industries.

Minnesota Department of Public Service, 1988
>   "A Survey of Minnesota Home Exterior Foundation Wall Insulation: Moisture Content and Thermal Performance," Unpublished Report, St. Paul, MN, November.

NAHB Research Center, 1993
>   "Assessment of Damage to Single-Family Homes Caused by Hurricanes Andrew and Iniki," Report prepared for U.S. Department of Housing and Urban

Development and Office of Policy Development and Research, Contract HC-5911, September.

National Association of Home Builders, 1963
Acoustical Manual. Washington, DC: National Association of Home Builders.

Portland Cement Association, 1995
"Insulating Concrete Forms: A Better Way to Build a Better Home," *Builder,* December.

Smith and Zungoli, 1995
Smith, B. C., and P. A. Zungoli, "Rigid Board Insulation in South Carolina: Its Impact on Damage, Inspection and Control of Termites," *Florida Entomologist,* Vol. 78, No. 3 (September), pp. 507–515.

Zollo, 1993
Zollo, Ronald, "Hurricane Andrew: August 24, 1992, Structural Performance of Buildings in Dade County, Florida," Technical Report No. CEN 93-1, University of Miami, Coral Gables, FL.

Part 2

Components

Although many of the parts and materials of ICF construction are familiar and used conventionally, others merit detailed description. Foremost are the systems themselves: their design, their components, and the details of their interconnection to create completed formwork. It is also important to understand the plastic foams that make up the bulk of the ICF units. Although most are now common in construction, ICFs rely on them more heavily and for some different tasks than conventional building does.

Chapter 2

ICF Systems

All ICFs are identical in principle: they consist of two parallel planes of a plastic foam material joined by regular crosspieces, to be stacked into the shape of walls and later filled with reinforced concrete. Beyond this, however, the various brands differ widely in the details of their shapes, their component parts, and their materials.

Categories of Systems

We categorize all ICF systems according to two characteristics that have broad implications: the form of the ICF unit and the form of the concrete in the finished wall.

The units exist in a variety of forms, which we group into *panel, plank,* and *block.* The key differences among them are their size, method of interconnection, and point of assembly. Figure 2-1 highlights the distinctive features of each.

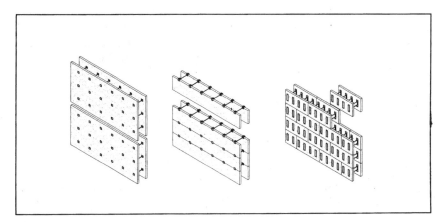

Figure 2-1 Panel, plank, and block systems.

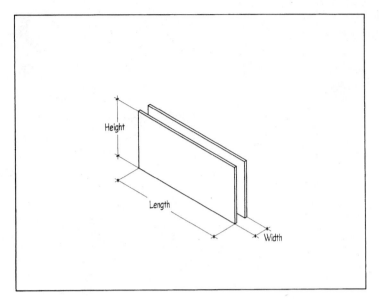

Figure 2-2 Dimension terminology conventions for ICF units.

Note that we report sizes according to the masonry standard of width×height×length (Figure 2-2). *Width* measures the distance between front and back surfaces. *Height* measures the smaller of the remaining dimensions and the one that is normally oriented vertically in a finished wall. *Length* measures the longer, normally horizontal, remaining dimension. The width of the unit is also the *thickness* of the final wall before adding finishes. When only two dimensions are reported, they are the height and length, unless specified otherwise.

Panel units are the largest, ranging from 1'3"×8'9" to 4'×12'. Their foam edges are flat or nearly so, and interconnection requires attachment of some separate connector or fastener, which is invariably made of another (nonfoam) material. The units are assembled (either in advance by the manufacturer or after delivery by the user) before setting. Plank units range from 1'×4' to 8"×8' to 1'×8'. Their unique feature is that they are assembled in place in the wall. Unlike panels and blocks, one does not assemble planks into complete units first and then set each unit into the formwork. Their name derives from the way they arrive on the job site. The long faces of foam are shipped as separate pieces (thus resembling wooden planks) and outfitted with crosspieces as part of the wall setting sequence. The edges of most planks are also flat, and interconnection is again by connectors or fasteners made of some other material.

Block units are the smallest: from 8"×1'4" to 1'4"×4'. They are molded with special edges that interconnect the blocks, much like plastic children's blocks snap together. The exact interconnect pattern varies from brand to brand. The most common are tongue and groove, interlocking teeth or nubs, or raised squares. Like panels, blocks are either preassembled in the factory or assembled by the user before setting.

The cavities inside the units also have various shapes. And since the concrete poured into them assumes their form, the shape of the resulting concrete wall varies correspondingly. The various final concrete forms fall into three categories: *flat, grid,* and *post and beam.* These are depicted in Figure 2-3. Flat concrete is of constant thickness, just like the concrete resulting from conventional plywood or metal formwork. Technically it is interrupted by periodic steel or plastic ties, just as conventional forms leave steel snap ties embedded in the concrete. But these comprise such a small percentage of the wall's surface area that they are neglected in most design decisions.

Grid concrete resembles a breakfast waffle, or a window screen made of thick wire. Cylindrical horizontal and vertical concrete members intersect, leaving thinner patches of concrete in between. The thin areas are interrupted by ties, technically leaving a break in the concrete. These are usually small again, but in some cases a thick tie leaves a larger interruption (up to 7 inches in diameter), as depicted in Figure 2-4. Sometimes grids with only small, negligible breaks in the concrete are called *uninterrupted grid* or *waffle*; grids with larger breaks are called *interrupted grid* or *screen*.

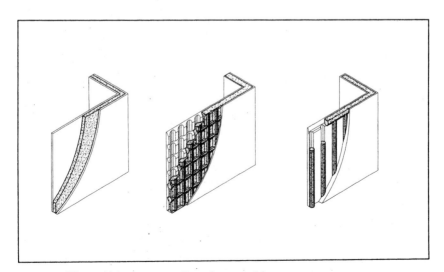

Figure 2-3 Flat, grid (uninterrupted), and post-and-beam systems.

32 Components

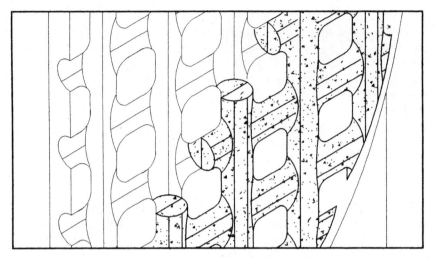

Figure 2-4 Cutaway view of a wall created with an interrupted grid system.

Post-and-beam concrete consists of more widely and variably spaced horizontal and vertical members. The name comes from the similarity of the concrete pattern to the structure of wooden post-and-beam walls. The distance between the concrete members can be varied to meet different structural requirements. One vertical "post" every 4 feet and one horizontal "beam" every 8 is the widest spacing commonly encountered in residential walls.

By matching each of the three unit forms with each of the three concrete (or cavity) forms, we can generate nine (3×3) different categories of ICFs and place all existing brands into one of these (Figure 2-5). Currently there are no brands in two of the nine categories (grid plank and post-and-beam plank). We therefore neglect these in the remainder of this book. Cutaway diagrams of systems from the other seven ICF categories (flat panel, grid panel, post-and-beam panel, flat plank, flat block, grid block, post-and-beam block) appear in Figures 2-6 through 2-12.

Formwork Components

Taken individually, ICF units are sometimes referred to simply as *forms*. Taken collectively, units assembled to form a wall, wall section, or set of walls are called *formwork*. Most ICF systems consist of more than one type of ICF unit. When they do, the most common unit, designed to be placed in a flat portion of the wall, is generally referred to as a *standard* or *stretcher* unit.

ICF Systems 33

	Form of ICF unit		
Form of cavities	Panel	Plank	Block
Flat	Lite-Form (preassembled) R-Forms	Diamond Snap-Form Lite-Form (unassembled) Polycrete Quad-Lock	Blue Maxx Fold-Form GREENBLOCK SmartBlock VWF
Grid	ENER-GRID RASTRA		I.C.E. Block Insulform Modu-Lock Polysteel Reddi-Form REWARD SmartBlock SF-10 Therm-O-Wall
Post & beam	Amhome ThermoFormed		ENERGYLOCK Featherlite Keeva

Figure 2-5 Categorization of available ICF systems.

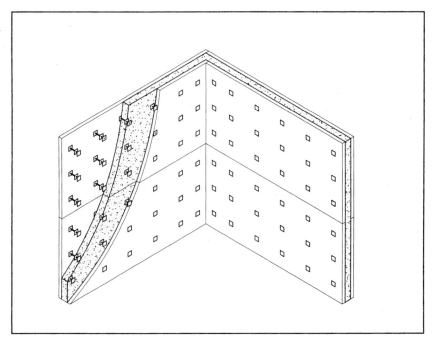

Figure 2-6 Cutaway view of a wall created with a flat panel system.

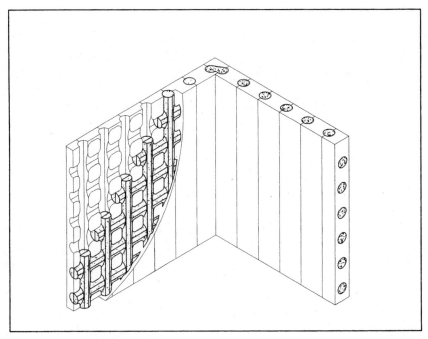

Figure 2-7 Cutaway view of a wall created with a grid panel system interrupted grid.

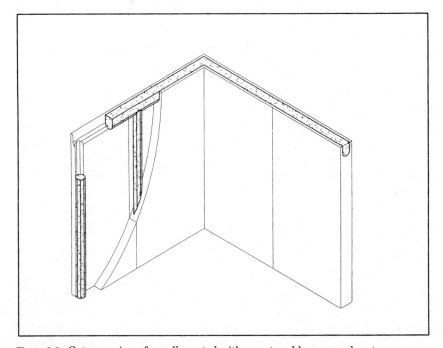

Figure 2-8 Cutaway view of a wall created with a post-and-beam panel system.

ICF Systems   35

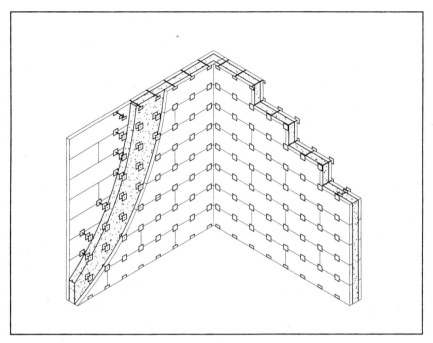

Figure 2-9 Cutaway view of a wall created with a flat plank system.

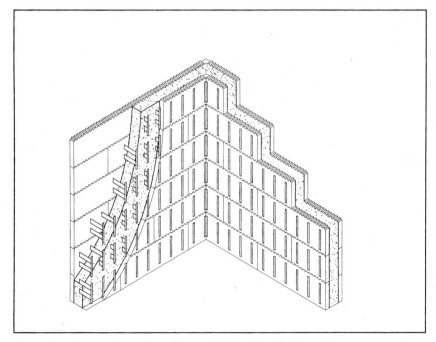

Figure 2-10 Cutaway view of a wall created with a flat block system.

36 Components

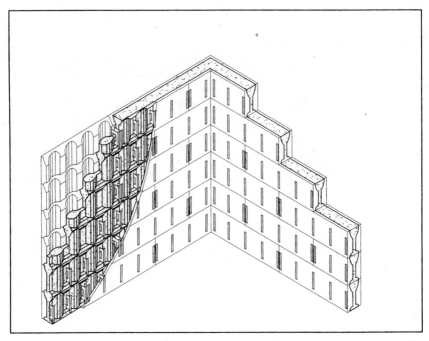

Figure 2-11 Cutaway view of a wall created with a grid block system (interrupted grid).

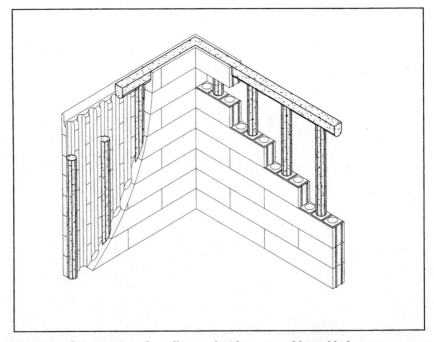

Figure 2-12 Cutaway view of a wall created with a post-and-beam block system.

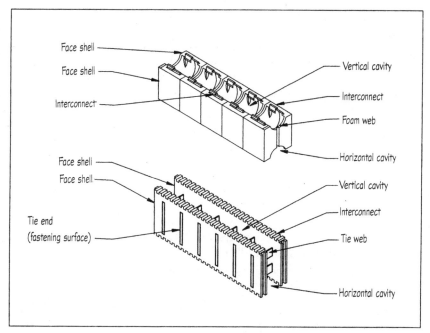

Figure 2-13 Parts of ICF units.

Parts of stretcher units

Figure 2-13 identifies the parts of stretcher units that are commonly distinguished. Terminology derives mostly from similar objects in conventional masonry and concrete form construction. The front and back planes of foam are sometimes called *face shells*. The surfaces of the face shells are the *faces,* or simply the front and back *surfaces*. The crosspieces connecting them are *ties*. When the ties are made of steel or solid plastic, they typically have a flat section at each end (front and back). These may be exposed or they may be hidden, embedded a fraction of an inch beneath the surfaces. They are called *tie ends* or *tie heads*. The remaining part of the tie connecting the face shells is called the *web*. When the unit, including the ties, is made entirely of foam, the ties have no heads per se, and they are often referred to simply as *foam webs* or *webs*.

The components used to connect two units to one another are referred to collectively as a system's *interconnect*. In block units the connection is by means of interlocking male and female shapes formed along the edges of the units. In these cases the interconnect is often referred to with more descriptive terms, such as tongue and groove or teeth. Some of these are continuously adjustable along the horizontal dimension (such as tongue and groove) and others are adjustable in fine increments (such as teeth). But many grid and post-and-beam block systems have an interconnect that offsets blocks on adjacent courses by exact increments of a vertical cavity. This enforces a strict alignment of vertical cavities, at the loss of some flexibility.

Interconnects used in panel and plank systems include one or more of the following: ties that bridge the seam between adjacent units; plastic channel that bridges the seam; wire or string wrapped around tie ends of adjacent units and across the seam; an adhesive applied along the units' edges.

In the field, sometimes a second interconnect is used to reinforce the first at potential weak points in the formwork. Specifically, interconnects that do not include wire or adhesives are sometimes supplemented with one of these or with an adhesive tape.

The units of many systems include a so-called *fastening surface*. This is a portion of the unit's face shell that contains some nonfoam material capable of taking and holding a standard fastener, such as a nail or screw. The primary uses of fastening surfaces are in the attachment of interior wallboard, exterior finishes, lightweight fixtures, and (temporarily, during wall construction) wall bracing. When using systems without such surfaces, these connections are made instead with adhesives to the formwork surface, connectors preinstalled before the pour and cast into the concrete, or concrete connectors shot or drilled into the concrete after hardening.

Fastening surfaces differ in the secondary functions they serve, material, and orientation. Most fastening surfaces are simply the units' tie ends. In block systems the tie ends are made of plastic or steel and form vertical strips from $\frac{3}{4}$ to 2 inches wide, spaced between each vertical cavity. When the units are correctly stacked, the tie ends align, providing continuous vertical fastening every 5, 8, 12, 16, or 24 inches on center, much like the studs of a frame wall. Some all-foam block systems sell, as a variant on their standard product, a block with similarly positioned plastic or wooden strips embedded in the face shells. Most plank and panel systems use plastic ties with rectangular ends approximately 3×3 inches, which provide fastening in an 8-×8-, 12-×12-, or 16-×16-inch grid pattern.

The plastic or steel channel interconnect used by certain panel and plank systems can serve as a fastening surface of about a 2-inch width. Depending on the system, they are usually horizontal and spaced anywhere from 1 to 4 feet.

Several all-foam systems will preinstall wooden strips in their units as an option. One provides plastic strips that can be inserted into slots in the formwork in the field. Note also that many manufacturers offer steel plates or plastic strips that attach anywhere onto the formwork after it is up. While these are not integral to the units, they serve the same purpose as a fastening surface.

Wooden strips provide the widest range of fastening options, since they can hold most common nails and screws. Solid plastic surfaces can accept most screws, as well as nails with high pullout (tensile) strengths, such as ringed or hot-dipped galvanized nails. Sheet metal fastening surfaces work with any screw that has a self-tapping ability. This includes common construction and wallboard screws, which are by far the most prevalent varieties on the typical residential job site.

Cavities and concrete sections

The cavities inside a unit or assembled formwork are sometimes collectively referred to as the *core*. In grid and post-and-beam systems, separate *vertical cavities* and *horizontal cavities* are discernible and are sometimes referred to by these names (Figure 2-14). Vertical cavities are also sometimes called *post cavities* or *cylinders*; horizontal cavities may be called *beam cavities*.

In flat systems the analogous vertical and horizontal spaces between ties will occasionally also be referred to as vertical and horizon-

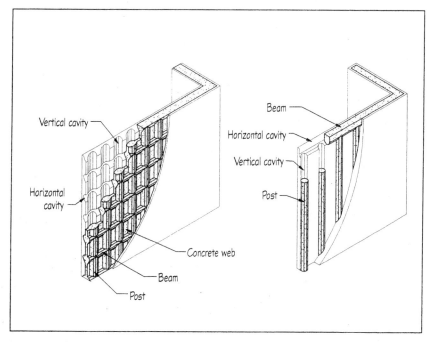

Figure 2-14 Types of ICF cavities and concrete members.

tal cavities. However, this is primarily a matter of convenience for identifying potential locations for housing rebar. These "cavities" are not distinguished by their shape from any other section of the core.

The members made of concrete that ultimately fill the vertical and horizontal cavities are sometimes referred to, respectively, as the *posts* and *beams* of the structure. In uninterrupted (waffle) grids, the thin patches of concrete between these members are called *concrete webs* (not to be confused with tie webs).

Specialty Units and Formwork Accessories

In addition to a stretcher unit, most systems also include special units to accomplish particular purposes. Most of these purposes can also be accomplished by appropriately modifying a stretcher unit. The specialty units, in effect, move that operation from the field to the factory. Various accessory parts are also available to adapt the formwork to specific uses.

Corners

In post-and-beam systems, corners are formed simply by placing a stretcher unit in the corner and abutting another against its side, as is

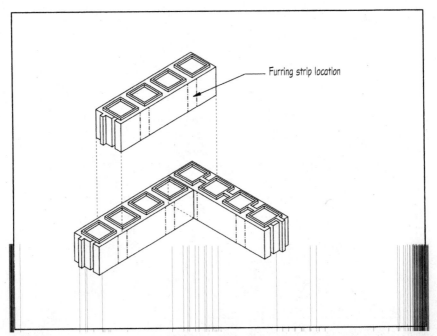

Figure 2-15 Abutting stretchers to form a corner.

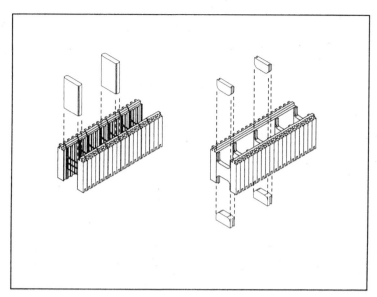

Figure 2-16 Inserting stops to form end blocks and stop horizontal concrete flow.

customary with masonry (Figure 2-15). In all other systems, unmodified stretchers cannot serve as corner units because their ends are open.

Some block systems provide foam *stops* (sometimes called *end pieces*), which are inserted into the openings at the end of a standard unit (Figure 2-16). Once capped, the stretcher becomes a so-called *end block* and can be placed in the corner like a post-and-beam block.

Corners for most flat panel and plank systems are field-created by cutting the inside face shells of two units short so that they butt to form a right angle (Figure 2-17). The units are secured with a variant of the system's interconnect (special plastic channel, or ties specially designed or arranged for a corner).

Except for grid blocks that use stops, all the grid systems and some flat plank systems depend on one or more of four distinct types of corner units: *user-cut, precut, preformed,* or *hinged*. It is always possible for the user to create a corner unit (user-cut) by miter cutting stretchers and adhering them into an angle (Figure 2-18). But some sellers also do the necessary cutting for 90° corners in advance (precut). The buyer is then responsible for assembling the precut halves, which is usually done with glue or tape.

Under most systems, the construction of either user-cut or precut units is done so as to produce two variants of the corner unit: one that is longer in both directions and one that is shorter. These allow installers to stagger the vertical joints between units up the wall, like

42 Components

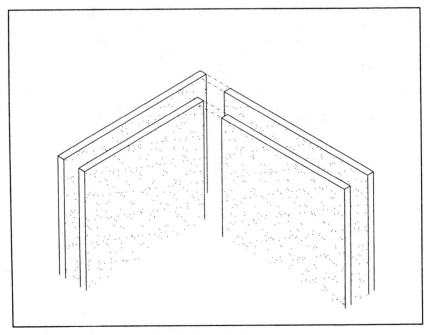

Figure 2-17 Alignment of panel or plank face shells to form a corner.

the running bond pattern common to masonry walls. The two types of corner units are sometimes referred to as *longs* and *shorts*.

Other sellers offer preformed corner units for right angles. As the name implies, these consist of L-shaped "planks" cut from a single block of foam, or 90° angle blocks molded as one piece in the factory. They typically come in two variants as well: either longs and shorts or a right-handed and a left-handed unit, as depicted in Figure 2-19. The two corner versions again allow one to stagger vertical joints from course to course.

Hinge units get their name because they can be rotated to form any angle over a wide range (Figure 2-20).

Some sellers without hinge units offer either precut or premolded units to form corners of 45°. If appropriate special units to form a particular angle are not available, the field cutting procedures described earlier can always be adjusted to do so.

Ties

Most plank systems include ties different from their standard ones. Since standard ties are designed to overlap the horizontal joints between planks, they would protrude inappropriately from the bottom of the first

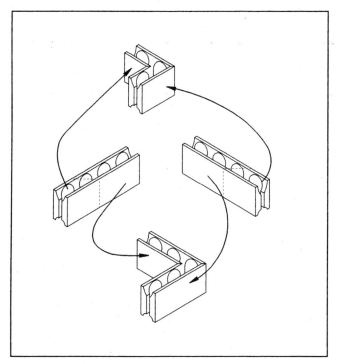

Figure 2-18 Cutting and gluing two stretchers (center) to form short (top) and long (bottom) corner units.

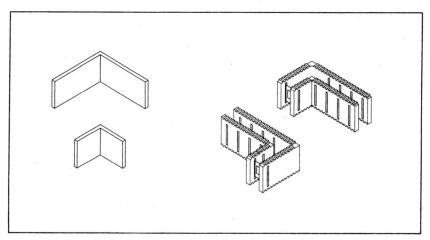

Figure 2-19 Preformed corners. Outside (top left) and inside (bottom left) face shells for a plank system, and left- and right-handed corner blocks (right) for a block system.

44 Components

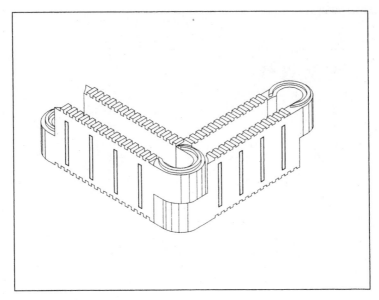

Figure 2-20 Hinge corner units.

course and above the top of the final course. Instead one installs *half-ties,* designed to sit flush with the edge of the plank (Figure 2-21). To distinguish them, the standard ties are sometimes called *full ties.*

Some plank systems without a preformed corner hold the field-assembled corners together with an assembly of standard ties, but others use a separate *corner tie,* also called a *full corner tie.* Half corner ties are also available for the tops and bottoms of the wall.

Lintel blocks

Post-and-beam systems require a continuous horizontal member (the *beam*) to tie the posts together structurally. Depending on loading, these beams are generally placed every 4 or 8 feet. In addition, extra lintels of reinforced concrete are necessary above some openings. Post-and-beam panel systems provide for such beams and lintels with horizontal cavities formed in the standard panel units. With post-and-beam blocks, the stretcher provides cavities only for vertical posts. Thus a special unit is necessary with cut-down webs to allow concrete to flow horizontally.

Following masonry terminology, these blocks are called *lintel, bond-beam,* or *U blocks.* It is possible to field-produce them by cutting the webs of stretchers. But most manufacturers of post-and-beam blocks premold a special unit, as depicted in Figure 2-22.

ICF Systems 45

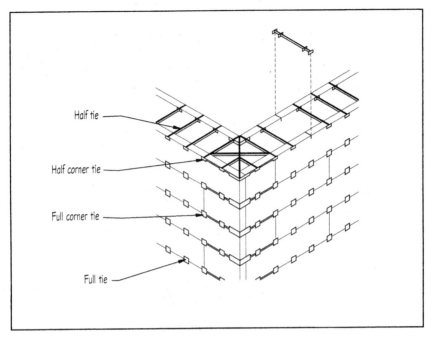

Figure 2-21 Ties of a plank system.

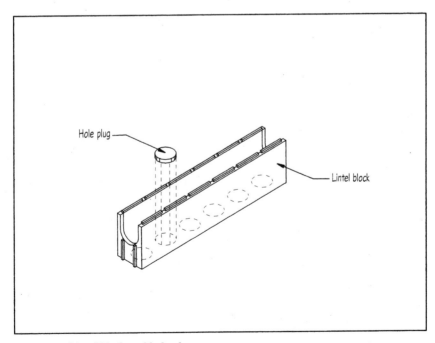

Figure 2-22 Lintel block and hole plug.

46 Components

A couple of other accessories are sometimes also provided. One is the *hole plug*. Most lintel blocks include holes at the bottom through which the concrete flows to fill the post cavities below. But most post-and-beam walls are designed with post spacing less frequent than at every cavity, so concrete must be prevented from entering some of them. Hole plugs are foam inserts placed into lintel block holes to seal over select cavities. This is analogous to the use of mason's felt to block the flow of grout into select cells of a reinforced concrete block wall. A second accessory is an end piece. Placing this in the end of a lintel block converts it to an end or corner unit, just as end pieces convert stretchers of some non-post-and-beam systems.

Some manufacturers of flat blocks also offer lintel units, usually with no holes at the bottom. These are not strictly required for their systems, but can be useful in particular situations.

Miscellaneous

A few manufacturers offer a *brick ledge unit*. Such units flare out on one side, as in Figure 2-23. Once filled, the concrete in the flared section of the formwork creates a ledge onto which to set a brick or other masonry veneer. Other manufacturers provide details for ways to modify standard units to provide the same function. The modification

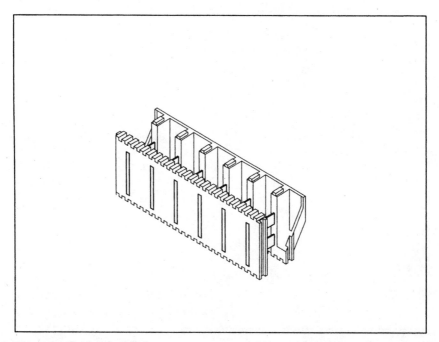

Figure 2-23 Brick ledge block.

may involve a reusable metal sheet that projects much like the flared foam face of the brick ledge unit.

As noted previously, some flat block manufacturers offer a foam stop to prevent concrete from flowing horizontally (see Figure 2-16). It can be placed inside the cavity at numerous points along the length of the block. Other than at corners, its primary use is to form the ends of the walls to either side of a window or door opening.

Bracing and Scaffolding Systems

ICF formwork requires varying amounts of supplemental bracing. Since the ties are generally sufficient to hold the formwork together against the force of the concrete during a pour, most of the bracing is not designed to "brace" in that sense. Rather, most ICF bracing serves to align the formwork and hold it precisely plumb and square. In addition, some type of scaffolding is usually necessary to elevate the crew for setting forms at higher levels and for placing the concrete.

Ordinary dimension lumber is frequently used for bracing (and then reused for other parts of the house), and common scaffolding equipment will give workers access to the top of the wall. However, a growing number of manufacturers offer special bracing or combined bracing and scaffolding systems.

The components of most of these systems consist of a set of standards that are set alongside the formwork (Figure 2-24). A standard's vertical post is flush with the wall, and a *kicker* extends diagonally to the ground. The standard and the kicker are steel channel, steel tubing, or 2×4 lumber. The standard attaches to the formwork, often at a fastening surface. The kicker is often adjustable so that workers can extend it as needed to plumb the wall. Also attached to the post is a bracket that holds scaffold planking.

OSHA regulations require that construction scaffolding have guardrails and toe boards on open sides and ends of all platforms more than 10 feet above the ground or floor, and in some cases when the platform is lower (National Archives and Records Administration, 1995). Although platforms are rarely higher than 10 feet in residential work, many of the ICF bracing and scaffolding systems have guardrails and toe boards (see Figure 2-24).

Available Systems

Table 2-1 lists the major characteristics of ICF systems sold in the United States. All of these are proprietary. Most of the systems undergo refinement in their design periodically. In addition, the manufacturers can differ in organizational characteristics (e.g., areas of ge-

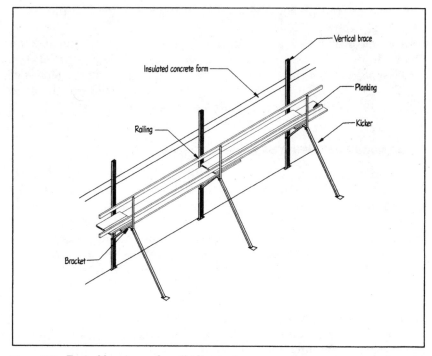

Figure 2-24 Typical bracing and scaffolding system.

ographic distribution, type and level of technical support offered) not covered in this book, and these things change over time as well. For complete and current information, always consult the manufacturer and the manufacturer's literature directly. The Directory of Product and Information Sources in Appendix C contains contact information. Note that systems sold in Canada only are not covered in detail in the book but appear in the directory.

Flat panel systems

Lite-Form is originally a flat plank system, but its manufacturer will preassemble planks into larger units, which more closely resemble panels. Preassembled Lite-Form is interconnected with a combination of the standard ties (from the plank version) and string or wire looped around the ties of adjacent panels (Figure 2-25). It shares many special components and construction details with the plank system (described in later sections). As an option it is also available with special hinged ties, which allow collapsing the unit so that it takes less space for shipping and storage (Figure 2-26). In this configuration it is sometimes referred to as Folding Lite-Form.

TABLE 2-1 Characteristics of Available ICF Systems

System	Dimensions of standard units	Primary material	Ties	Inter-connect	Integral fastening surface Identity	Integral fastening surface Orientation	Corners	Other specialty units	Bracing/scaffolding systems
Flat panel systems:									
Lite-Form (pre-assembled)	1'×2'×8' 1'×4'×8' or to order	Optional[a]	Plastic	Ties	Tie ends	Rectangles 8" oc, vertically & horizontally	User-cut	None	None
R-FORMS[b]	8"×4'×8' 10"×4'×8' 1'×4'×8' 14"×4'×8'	Optional[c]	Plastic	Plastic channel	Tie ends	Rectangles 16" oc, vertically & horizontally	User-cut	None	Bracing system
Grid panel systems:									
ENER-GRID	8"×1'3"×10' 8"×2'6"×10' 10"×1'3"×10' 10"×2'6"×10' 1'×1'3"×10' 1'×2'6"×10'	Foam/cement composite	Foam/cement webs	Limited adhesive	None	—	Abutted	None	None

[a] Manufacturer will cut various types of 2-inch foam sheet into units, as specified by buyer.
[b] Lengths of up to 30 feet are also possible.
[c] Units are user-assembled of 2-inch EPS or XPS foam sheet, independently purchased.
[d] Units are user-cut from 9⅜-inch foam sheets, independently purchased, but specified by Amhome to be expanded polystyrene.
[e] Wood straps are field-installed at optional spacing, but in practice generally 16" oc vertically (analogous to studs of frame construction).
[f] Some plants that produce the system will supply non-90° corners or curved units on request.

TABLE 2-1 Characteristics of Available ICF Systems *(Continued)*

System	Dimensions of standard units	Primary material	Ties	Interconnect	Integral fastening surface		Corners	Other specialty units	Bracing/scaffolding systems
					Identity	Orientation			
Grid panel systems *(Cont.)*:									
RASTRA	8"×1'3"×10' 8"×2'6"×10' 10"×1'3"×10' 10"×2'6"×10' 1'×1'3"×10' 1'×2'6"×10' 1'2"×1'3"×10' 1'2"×2'6"×10'	Foam/cement composite	Foam/cement webs	Limited adhesive	None	—	Abutted	None	None
Post-and-beam panel systems:									
Amhome	9⅜"×4'×8'	Expanded polystyrene[d]	Foam webs	Adhesive	Wood strap	Vertical strips 16" oc[e]	Abutted	None	None
ThermoFormed	8"×4'×8'	Expanded polystyrene	Foam webs	Steel channel	Steel channel interconnect	Horizontal strips 4' oc	Abutted	None	None
Flat plank systems:									
Diamond Snap-Form	8"×1'×8' 10"×1'×8' 12"×1'×8' 12"×1'×8'	Expanded polystyrene	Plastic	Ties	Tie ends	Diamonds 12" oc, vertically & horizontally	Preformed	Plant-dependent[f]	None
Lite-Form (nonassembled)	1'×8"×8'	Optional[a]	Plastic	Ties	Tie ends	Rectangles 8" oc, vertically & horizontally	User-cut	None	None

System	Unit dimensions	Foam material	Tie material	Tie connection	Tie ends	Web/tie spacing	Corners	Special units	Bracing
Polycrete	10⅝"×1'×8' 11"×1'×8' 1⅝"×1'×8' 11"×1'×8' 12⅝"×1'×8' 1'3"×1'×8' 14⅝"×1'×8'	Expanded polystyrene	Steel	Integral plastic channel	Integral plastic channel	Strips 12" oc, horizontally	User-cut	None	Bracing/scaffolding system
Quad-Lock	8⅛"×1'×4' 10⅛"×1'×4' 12⅛"×1'×4' 12⅛"×1'×4'	Expanded polystyrene	Plastic	Ties and integral nubs	Tie ends	Rectangles 12" oc, vertically & horizontally	User-cut	None	None
Flat block systems:									
Blue Maxx	11.5"×16.75"×4' 12.625"×16.75"×4'	Expanded polystyrene	Plastic	Integral teeth	Tie ends	Vertical strips 8" oc	Preformed	Hinge, brick ledge	Bracing/scaffolding system
Fold-Form	8"×1'×4' 1'×1'×4' 10"×1'×4'	Expanded polystyrene	Plastic	Integral teeth	Tie ends	Rectangles 8" oc, vertically & horizontally	User-cut	None	None
GREENBLOCK	9.88"×9.88"×3'3.4" (250 mm × 250 mm × 1000 mm)	Expanded polystyrene	Plastic	Integral nubs	Tie ends	Vertical strips 5" oc	Preformed	45° corner, lintel block, brick ledge, radius block, end piece	Bracing/scaffolding system

[a] Manufacturer will cut various types of 2-inch foam sheet into units, as specified by buyer.
[b] Lengths of up to 30 feet are also possible.
[c] Units are user-assembled of 2-inch EPS or XPS foam sheet, independently purchased.
[d] Units are user-cut from 9⅜-inch foam sheets, independently purchased, but specified by Amhome to be expanded polystyrene.
[e] Wood straps are field-installed at optional spacing, but in practice generally 16" oc vertically (analogous to studs of frame construction).
[f] Some plants that produce the system will supply non-90° corners or curved units on request.

TABLE 2-1 Characteristics of Available ICF Systems (Continued)

System	Dimensions of standard units	Primary material	Ties	Interconnect	Integral fastening surface		Corners	Other specialty units	Bracing/scaffolding systems
					Identity	Orientation			
Flat box systems (Cont.):									
SmartBlock VWF	8"×1'1"×3'4" 10"×1'×3'4" 1'×1'×3'4" 1'2"×1'×3'4"	Expanded polystyrene	Plastic	Integral teeth	Tie ends	Vertical strips, optional centers	Stop	None	None
Grid block systems:									
I.C.E. Block	9"×14"×4' 11"×14"×4'	Expanded polystyrene	Steel or plastic[e]	Integral tongue & groove	Tie ends	Vertical strips 12" oc	User-cut or preformed[e]	None	None
Insulform	9.6"×14"×4'	Expanded polystyrene	Foam web	Integral teeth	Optional plastic strip	Horizontal strips 16" oc	Preformed	End	None
Modu-Lock	10"×10"×3'4"	Expanded polystyrene	Foam web	Integral teeth	None	—	Stop	None	None
Polysteel	9"×14"×4' 11"×14"×4'	Expanded polystyrene	Steel	Integral tongue & groove	Tie ends	Vertical stips 12" oc	Precut	None	None
Reddi-Form	9⅝"×1'×4'	Expanded polystyrene	Foam web	Integral teeth	Optional plastic strip	Horizontal strips 12" oc	Preformed	End, pilaster	None
REWARD	9"×14"×4' 11"×14"×4'	Expanded polystyrene	Plastic	Integral shiplap	Tie ends	Vertical strips 12" oc	Preformed	None	None

System	Size	Material	Web	Interlock	Reinforcement ties	Reinforcement placement	End treatment	Special shapes	Other
SmartBlock SF 10	10"×10"×3'4"	Expanded polystyrene	Foam web	Integral teeth	None	—	Stop	None	None
Therm-O-Wall	9"×1'4"×4' 11"×1'4"×4'	Expanded polystyrene	Plastic	Integral tongue & groove and adhesive	Tie ends	Vertical strips 12" oc	Preformed	None	None
Post-and-beam block systems:									
ENERGY LOCK	8"×8"×2'8"	Polyurethane	Foam web	Integral raised squares	Optional wood strip	Vertical strips 16" oc	Abutted stretchers	Lintel	None
Featherlite	8"×8"×1'4"	Polyurethane	Foam web	Integral raised squares	None	—	Abutted stretchers	None	None
KEEVA	8"×1'×2'8"	Expanded polystyrene	Foam web	Integral raised squares	None	—	Abutted stretchers	Lintel	None

[a] Manufacturer will cut various types of 2-inch foam sheet into units, as specified by buyer.
[b] Lengths of up to 30 feet are also possible.
[c] Units are user-assembled of 2-inch EPS or XPS foam sheet, independently purchased.
[d] Units are user-cut from 9⅜-inch foam sheets, independently purchased, but specified by Amhome to be expanded polystyrene.
[e] Wood straps are field-installed at optional spacing, but in practice generally 16" oc vertically (analogous to studs of frame construction).
[f] Some plants that produce the system will supply non-90° corners or curved units on request.

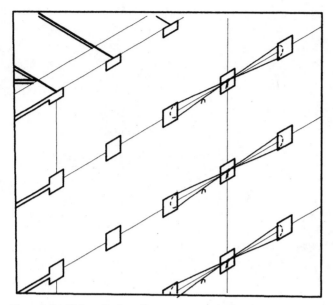

Figure 2-25 Tying plank units across a vertical seam.

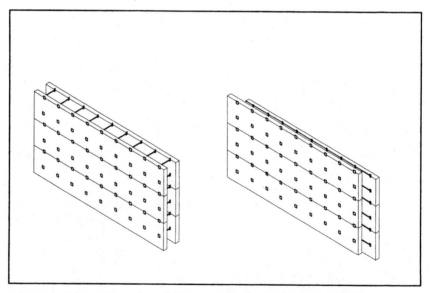

Figure 2-26 Folding Lite-Form unit fully expanded (left) and collapsed (right).

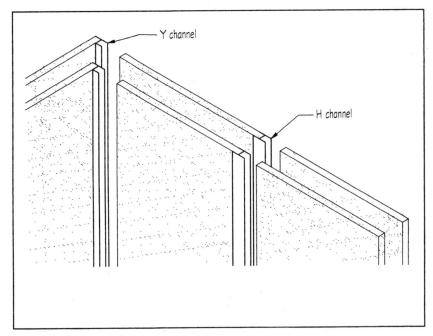

Figure 2-27 Joining adjacent R-FORMS panels with plastic channel.

R-FORMS is a user-assembled panel, depicted in Figure 2-27. The user selects and buys 2-inch foam sheet and inserts the R-FORMS ties. Interconnection is by means of plastic channel, available from the manufacturer: an H-channel for a straight (nonangled) connection and a Y-channel for 90° angles.

Grid panel systems

ENER-GRID and RASTRA are physically similar. The cavity geometry of each is an interrupted grid with wide webs. They have the unique attribute of being constructed out of a mixture of cement and foam beads, rather than pure foam. This makes the units relatively stiff and potentially more durable in some respects, in return for greater weight and a higher cost. Because the weight helps hold the units in place, interconnection is normally by intermittent gluing of the joints only (Figure 2-28).

Post-and-beam panel systems

The Amhome system is more broadly inclusive than the others, both in how it is marketed and what portions of the house it includes. The

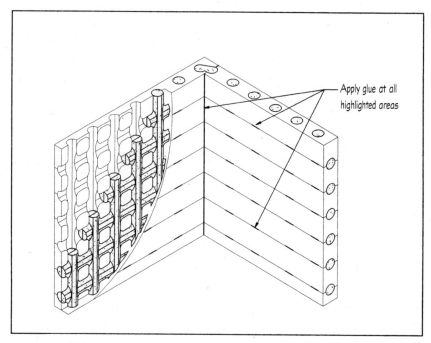

Figure 2-28 Gluing locations on grid panel systems.

parent corporation sells some complete precut Amhome kits. However, otherwise it does not sell any parts or units. It licenses the patented technology, special tools, and rights to build Amhomes. To build them a contractor must first enter into a licensing agreement with the company. The company provides the tools and instructions to make units from standard foam sheet. In addition to the wall units, the company provides specifications for constructing the roof according to its own original, superinsulating methods, and recommends details for some other parts of the house.

The Amhome panel units themselves are generally $9\frac{3}{8}"\times 4'\times 8'$, with one post and one beam per unit. In addition, slots may be cut into the foam and wooden or metal straps inserted to serve as a vertical fastening surface every 16 or 24 inches. See Figure 2-29 for details.

ThermoFormed Block Corporation licenses local businesspeople to fabricate units of its patented ThermoFormed system for their own use or for sale to the public. The licensees make the units from locally purchased foam sheet stock. As shown in Figure 2-30, the panel units consist of two 4×8 sheets connected with long foam webs glued into place during fabrication. The fabricators can include stops over select vertical cavities of the units, so that only every second or third verti-

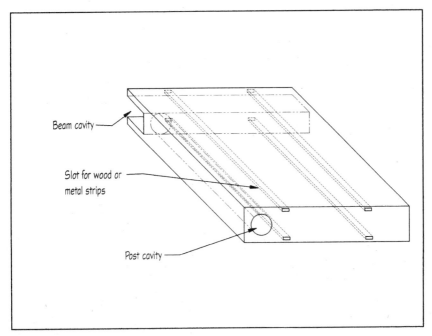

Figure 2-29 Cavities of Amhome unit.

cal cavity fills with concrete. Or they can install no stops, and the unit will be completely filled. Steel rails interconnect panels and also serve as a fastening surface.

Flat plank systems

Diamond Snap-Form was designed by Associated Foam Manufacturers (AFM) and is produced and sold by foam molding companies that are members. Diamond Snap-Form employs planks notched along the top and bottom edges. Ties fit into the notches. As the ties overlap the seams and grip each panel, they serve as the standard interconnect. Corners are preformed. See Figure 2-31 for details. Some companies that sell Diamond Snap-Form provide specialty units on request. These include non-90° corners and curved units.

Unassembled Lite-Form is similar to Diamond Snap-Form; the panels are narrower and the ties spaced more closely. Lite-Form uses corner ties like those in Figure 2-21 to connect standard planks into corners. Lite-Form also offers ties that allow easy stripping of the foam from one or both sides after pouring, if that is desirable.

Polycrete planks, rather than being cut from stock foam sheet, are molded with a T-shaped plastic channel protruding from the top edge,

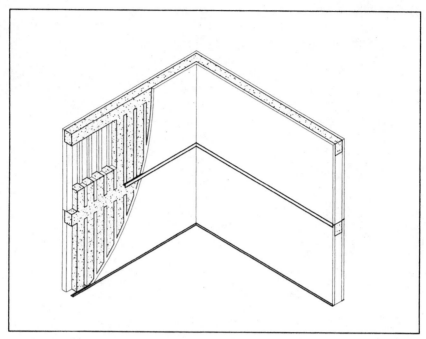

Figure 2-30 Cutaway view of a wall created with ThermoFormed panels.

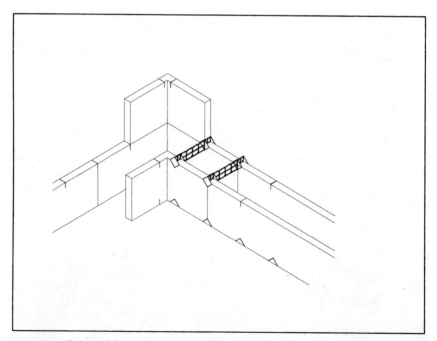

Figure 2-31 Diamond Snap-Form under assembly.

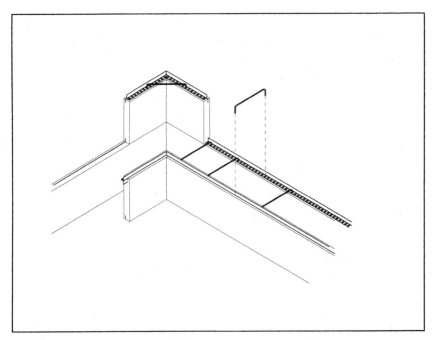

Figure 2-32 Polycrete under assembly.

as visible in Figure 2-32. The channel serves three purposes. It is a horizontal fastening surface, it is the "tongue" of the interconnect (the opposite edges of the planks have grooves into which it fits), and it holds the ties. The ties are steel rods with a right angle at each end, which workers push into preformed holes in the channel. Corners are constructed out of standard planks and a special arrangement of standard ties.

Quad-Lock planks are also molded, with slots in the edges into which ties fit. The edges have a molded interlock to reinforce the interconnection, much like block systems. Corners are of standard planks and a special arrangement of the ties. Figure 2-33 shows details.

Flat block systems

All of the flat block systems (depicted in Figure 2-34) have solid plastic ties. Blue Maxx has a wide range of specialty units and parts. Fold-Form, from the makers of Lite-Form, has hinged ties that allow the unit to collapse for shipping and storage (much like Folding Lite-Form). GREENBLOCK has face shells thicker on one side than on the other. The thicker side faces out, so there is a specific front-back orientation to the unit. GREENBLOCK has a wide range of special parts and units available.

60 Components

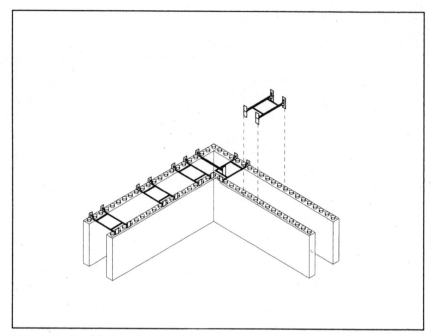

Figure 2-33 Quad-Lock under assembly.

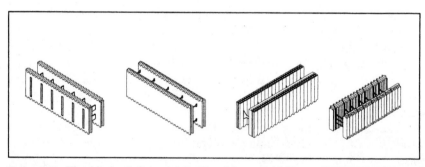

Figure 2-34 Flat blocks (left to right): Blue Maxx; Fold-Form; GREENBLOCK; and SmartBlock VWF.

SmartBlock VWF (variable-width form) differs from the other flat block systems in that the units come unassembled. The face shells are shipped as separate pieces, and the ties slide into grooves along their inside surfaces. Users can vary the spacing of the ties within the unit, and by using longer or shorter ties, they can vary the widths of the assembled units and the cavities inside them. An available foam stop closes off the ends of stretchers to form end blocks for corners.

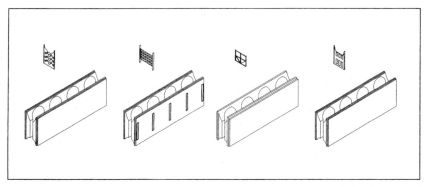

Figure 2-35 Uninterrupted grid blocks and their ties (left to right): I.C.E. Block and sheet steel tie (plastic also available); Polysteel and expanded lath steel tie; REWARD and plastic tie; and Therm-O-Wall and plastic tie.

Grid block systems

The four blocks with an uninterrupted grid (I.C.E. Block, Polysteel, REWARD, and Therm-O-Wall) are descendants from a single, earlier system. As a result they are similar in design, as depicted in Figure 2-35. All use a tongue-and-groove interconnect, except for REWARD, which uses a shiplap. All have units of the same overall dimensions, except that some brands come in two widths. Their cavity geometries are nearly identical. The ties differ somewhat in design and materials; some manufacturers offer steel, some rigid plastic, and some both. The systems also vary in the types of corner units available.

Organizationally, I.C.E. Block is unique. It is produced by several manufacturers that market jointly but serve separate regions of the country. All of these manufacturers' I.C.E. Block is primarily the same, except for some variations in tie design and materials, and in the types of corner units available.

The other four grid block systems employ all-foam blocks with interrupted grids. Insulform includes a stretcher, corner, and end units. Its wide-tooth interconnect allows only offsetting blocks on adjacent courses by exact increments of a vertical cavity. Available plastic strips can be preinserted into slots along the blocks' horizontal edges to serve as a fastening surface (Figure 2-36).

Reddi-Form includes a stretcher, corner, and end units, as well as a pilaster unit. The pilaster allows the formation of a continuous vertical post uninterrupted by any beams. Like Insulform, Reddi-Form has available plastic strips for optional insertion along the blocks' horizontal edges to serve as a fastening surface, and a wide-tooth interconnect allows only precise vertical cavity alignment.

Modu-Lock includes stretcher and premolded end blocks. Its inter-

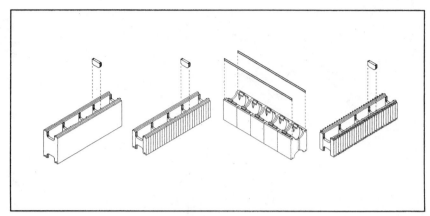

Figure 2-36 Interrupted grid blocks (left to right): Insulform, Modu-Lock, Reddi-Form and optional slide-in plastic fastening surface; SmartBlock SF 10 and foam stop.

connects offset blocks on adjacent courses by exact increments of a vertical cavity.

SmartBlock SF 10 consists of a stretcher plus foam stops to form corners and ends. SF 10's fine-tooth interconnect is compatible with that of the SmartBlock VWF product, although its width is fixed.

Post-and-beam block systems

The post-and-beam block systems all use a raised square integral interconnect that offsets blocks on adjacent courses by exact increments of a vertical cavity (Figure 2-37). They also have tongue-and-groove interconnects along their ends, although the configurations of these differ across brands.

Major differences are size, accessory parts, and material. ENERGY LOCK has the dimensions of a double-length CMU (8×8×32 inches) and provides hole plugs for lintel blocks. It is available with optional wooden straps (molded into the foam) to serve as fastening surfaces. Featherlite is sized like a standard CMU, and relies on the field personnel to use mason's felt or a similar product to cover unfilled post cavities. ENERGY LOCK and Featherlite are the only two ICFs with units made of a polyurethane, rather than a polystyrene or composite. (See discussion of foams in Chapter 3.) KEEVA comes in 4-foot-long units (8×8×48 inches) and also offers hole plugs.

ICF Systems 63

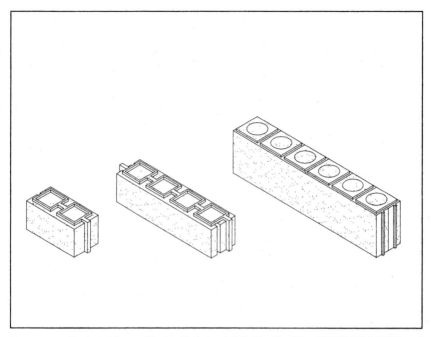

Figure 2-37 Post-and-beam blocks (left to right): Featherlite; ENERGY LOCK; and KEEVA.

Reference

National Archives and Records Administration, 1995
 Code of Federal Regulations, Vol. 29, Part 1926.451, Subpart L. Washington, DC: U.S. Government Printing Office.

Chapter

3

Plastic Foams

Plastic foams are the predominant material of ICF formwork and one of the major materials of the finished walls. Understanding their properties helps in specifying products properly and designing the rest of the structure.

Classification

Figure 3-1 relates the different classes of foams used in ICFs. Table 3-1 lists the class employed in each of the available systems.

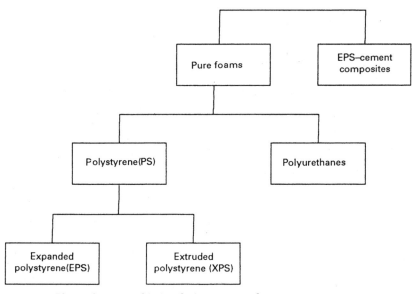

Figure 3-1 Plastic foams used in insulating concrete forms.

TABLE 3-1 Plastic Foams in ICF Systems

System	Foam class	Nominal density (lb/cu ft)*	ASTM type[†]
Flat panel systems:			
Lite-Form (preassembled)[‡]	XPS	2.0	IV
R-FORMS[‡]	EPS	2.0	
	XPS	1.7	IV
Grid panel systems:			
ENER-GRID	Composite	21	
RASTRA	Composite	21.2	
Post-and-beam panel systems:			
Amhome[‡]	EPS	1.0–1.5	
ThermoFormed[‡]	EPS	1.0	
Flat plank systems:			
Diamond Snap-Form	EPS	2.0	
Lite-Form (nonassembled)[‡]	XPS	1.8	IV
Polycrete	EPS	2.0	
Quad-Lock	EPS	2.0	
Flat block systems:			
Blue Maxx	EPS	1.5	II
Fold-Form	EPS	2.0	
GREENBLOCK	EPS	1.5	II
SmartBlock VWF	EPS	2.0	
Grid block systems:			
I.C.E. Block	EPS	1.5	II
Insulform	EPS	2.0	
Modu-Lock	EPS	1.5	
Polysteel	EPS	1.5	
Reddi-Form	EPS	2.0	
REWARD	EPS	1.5	
SmartBlock SF 10	EPS	1.5	
Therm-O-Wall	EPS	1.5–2.0	
Post-and-beam block systems:			
ENERGY LOCK	Polyurethane	2.0	
Featherlite	Polyurethane	2.0	
KEEVA	EPS	1.5–2.0	

*If reported.
[†]If reported. Applicable to EPS and XPS only.
[‡]Suggested specifications for the foam recommended by the manufacturer. Other sheet foams may also be used.

Polystyrenes

Most of the ICF systems use a form of polystyrene foam, either molded expanded (EPS) or extruded expanded (XPS). Both EPS and XPS are manufactured from polystyrene resin and a blowing agent (a gas that expands when heated) to create a rigid, closed-cell, cellular plastic foam. EPS manufacture uses heat and pressure to expand small polystyrene "beads." The final expansion occurs in a mold, where the beads fuse into a single mass. The mold may be shaped to form the EPS directly into the final ICF product (*shape-molded* in industry terminology) or to form it into a large block (*block-molded*) that is then cut into the ICF foam parts. All ICF blocks made from EPS and some planks are shape-molded. Some EPS planks and all the "boards" used in panel systems are cut from block-molded material.

XPS is produced in a continuous extrusion process, which results in a homogeneous cellular structure. XPS insulation for construction is available in the form of boards in various thicknesses.

By adjusting production variables and introducing additives, manufacturers of either type of foam can vary its properties to meet the specific requirements of a wide range of ICF applications. One pivotal property of either polystyrene foam is its density, measured in pounds per cubic foot of material. Altering a foam's density can change other important physical properties of the material, notably R-value and strength. However, whether and how much a particular attribute changes with density depends on the type of foam. Some properties that change with density in EPS are virtually constant in XPS (see "Properties").

All polystyrene foams approved for use in construction also contain additives that retard combustion of the material to comply with surface burning requirements of the model building codes. In the case of EPS, foam containing such additives is referred to as *modified* material.

Polyurethanes

A few systems employ a polyurethane foam. In current usage, the term polyurethane refers to a range of plastic foams that combine some form of isocyanate, some form of polyol, and a blowing agent. The first two chemicals react when combined to form polyurethane plastic. The blowing agent, caught in the middle of the reaction, froths the mixture with numerous tiny cells to produce a foam.

The blowing agents used have greater thermal resistance than ordinary air. Since they fill the cells of the foam, this raises the foam's R-value. Compared to the EPS-based materials described previously, polyurethanes generally have higher thermal resistance and a higher cost. Their precise properties vary somewhat with the exact formulation, however. They also vary with other parameters, notably density and age. Density can be varied much as it can in polystyrene foams, with similar results.

EPS-cement composites

A few systems use a material that combines EPS beads and portland cement. The mix also includes additives that improve adherence of the two main ingredients. The beads are conventional EPS beads, as described, which are expanded but not fused together. Instead, they are mixed with the cement, additives, and water to form an extremely lightweight insulating concrete. This goes into a mold, where it hardens to form the units.

The exact formulas for these materials are proprietary. Their manufacturers have given them various brand names, such as PaFcon and Thastyron. The composites enjoy greater strength and durability than many conventional foams, in return for greater weight and, apparently, higher cost.

Properties

Table 3-2 contains typical values of important properties for the foams commonly used in ICFs. The exact parameters of a particular class of foam varies with its raw materials, equipment, and production methods. Therefore the literature of many ICF manufacturers lists not only the type of foam used, but also data on its key properties.

It is important for designers to recognize that the formulation, or even the class of foam, used by an individual manufacturer can change over time. If precise property values are critical, it is advisable to get the most recent product data from the manufacturer. (See Appendix C for contact information.)

Foam properties are relevant to deciding which ICF system best fits a particular application. However, the systems that construct their ICF units of sheet stock allow the user to separate the choice of a foam from the choice of an ICF, at least partially. Although the manufacturers of these systems typically recommend a particular foam, the designer may substitute any of several that are available in sheet form.

Polystyrenes

ASTM standard C578-95 (ASTM, 1995) is the thermal insulation specification for polystyrene foams. It sets forth 11 sets of minimum performance specifications, called "types." Foams meeting a certain set of specifications are designated type I, those meeting a certain other set, type II, and so on. The standard does not differentiate EPS and XPS.

Plastic Foams

TABLE 3-2 Typical Properties of Plastic Foams Used in ICFs*

Property[a]	EPS	XPS	Composite	Polyurethane
Density (lb/cu ft)[b] [kg/m^3]	1.35–1.8 [24.0–29.0]	1.6–1.8 [26.0–29.0]	21 [336.4]	2.0 [32.0]
R per inch[c] [m$^2 \cdot$ °C/W]	4.17–4.35 [0.73–0.76]	5.0 [0.88]	3.0 [0.53]	5.9[d] [1.04]
Strength (lb/sq in) [kg/m2]				
Compressive[e]	15–33 [73–160]	25–40 [122–195]	72 [352]	30 [146]
Flexural	40–75 [195–366]	50–60 [244–293]	75 [385]	
Tensile	88–27 [88–132]	45–75 [220–366]	42 [205]	30 [146]
Shear	26–37 [126–180]	30–35 [146–170]		35 [171]
Water vapor permeance per inch (perms) [g/(Pa $\cdot$ s $\cdot$ m^2)]	1.0–3.5 [0.6×10^{-7}– 2.0×10^{-7}]	1.1 [6.3×10^{-8}]		2.0 [1.1×10^{-7}]
Water absorption (%)[f]	<3.0	<0.3		2.0
Dimensional stability (%)	<2.0	<2.0		7.0
Flame spread	10	5		20
Smoke developed[g]	125	165		250
Cost ($/bd ft)[h] [$/m]	0.17 [0.05]	0.35 [0.11]		0.70 [0.21]

*Metric values are given in brackets.
[a]All properties will vary with details of the foam's formulation and environmental conditions. See text for discussion.
[b]Values of most other properties vary sharply with density. Foams of other densities will have values different from those listed here.
[c]At 75°F ambient temperature.
[d]For material that is fully aged and not faced with an impermeable membrane.
[e]At 10 percent deformation.
[f]By total immersion. Percentage absorption by volume.
[g]For 2-inch thick material.
[h]Retail price in small quantities, 1-inch-thick sheet material.

But generally speaking, there are EPS foams available that meet the specifications of types I, II, VIII, IX, or XI. There are XPS foams available that meet the specifications of types IV, V, VI, VII, or X. Table 3-3 reproduces the specification sets of the standard. The National Building Code of Canada requires polystyrene insulation to conform to a different standard, CAN/CGSB 51.20-M87 (see Appendix B).

The ASTM classification system provides a useful shorthand for manufacturers and designers. The manufacturer can convey a whole range of properties with merely the "type" of the foam. The designer can specify a set of properties with the same brevity.

Other names have arisen for some of the ASTM types in common usage. Type I foam is referred to as "nominal one-pound density" EPS. Although the minimum density for type I is 0.90 lb/cu ft, in practice this is generally met by foams that are manufactured to be 1.0 lb/cu ft and are advertised as such by their sellers. Since manufacturing processes may vary, densities of individual type I pieces may be slightly above or below 1.0 lb/cu ft, but by the standard they must always exceed 0.90 lb/cu ft. Analogously, types II and X are "nominal one-and-a-half pound" EPS and XPS, respectively, and types IX and VI are "nominal two-pound" EPS and XPS.

Note that most other specifications vary with density. In this matter, the standard reflects the close relationship between density and other physical properties of the foam. Figures 3-2 and 3-4 through 3-7 depict these relationships graphically. In fact, in the past designers have specified their polystyrene foam by stating only the required class (EPS or XPS) and the density. However, since other factors influence foam properties, density alone is a less reliable guarantor than ASTM type.

Thermal resistance. Figure 3-2 charts the increases in specified thermal resistance (R-value) that accompany higher specified densities. Note, however, that this relationship is flat for XPS. The R-value of XPS actually varies with density below 0.3 lb/cu ft. But such low densities are not common. Note also that for EPS the relationship experiences diminishing returns. For example, beyond 2 lb/cu ft the R-value increases little with greater density. The R-value also rises with density for fiberglass insulation, whose numbers are included in the figure for comparison.

Thermal resistance also varies with the foam's temperature (as discussed in Chapter 1) as well as with its density. Polystyrene foam retards heat flow slightly more effectively when it is cold, regardless of the class or type. Figure 3-3 illustrates the relationship for a couple of representative formulations and, for comparison, batt fiberglass.

TABLE 3-3 Specifications of ASTM C578 for Polystyrene Foams

Property	Classification									
	Type XI	Type I	Type VIII	Type II	Type X	Type IV	Type IX	Type VI	Type VII	Type V
Density, min (lb/cu ft)	0.70	0.90	1.15	1.35	1.30	1.60	1.80	1.80	2.20	3.00
Thermal resistance of 1.00 inch, min (°F · sq ft · h/Btu) at mean temperature:										
25°F	3.45	4.20	4.40	4.60	5.60	5.60	4.80	5.60	5.60	5.60
40°F	3.30	4.00	4.20	4.40	5.40	5.40	4.60	5.40	5.40	5.40
75°F	3.10	3.60	3.80	4.00	5.00	5.00	4.20	5.00	5.00	5.00
110°F	2.90	3.25	3.45	3.65	4.65	4.65	3.85	4.65	4.65	4.65
Compressive resistance at yield or 10 percent deformation, min (psi)	5.0	10.0	13.0	15.0	15.0	25.0	25.0	40.0	60.0	100.0
Flexural strength, min (psi)	10.0	25.0	30.0	40.0	40.0	50.0	50.0	60.0	75.0	100.0
Water vapor permeance of 1.0 inch, max (perm)	5.0	5.0	3.5	3.5	1.1	1.1	2.0	1.1	1.1	1.1
Water absorption by total immersion, max (vol-%)	4.0	4.0	3.0	3.0	0.3	0.3	2.0	0.3	0.3	0.3
Dimensional stability (change in dimensions), max (%)	2.0	2.0	2.0	2.0	2.0	2.0	2.0	2.0	2.0	2.0
Oxygen index, min (vol-%)	24.0	24.0	24.0	24.0	24.0	24.0	24.0	24.0	24.0	24.0

SOURCE: American Society for Testing and Materials (1995). Reprinted with permission of ASTM.

Components

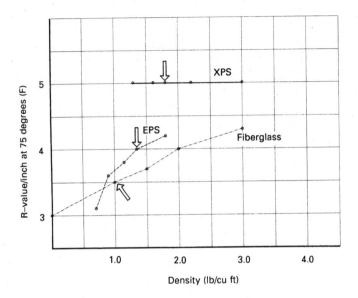

Note: ⇧ Arrow denotes most commonly used density.

Note: EPS and XPS data points are for specified minimum R-values. Actual R-values may be higher.

Figure 3-2 Specified R-values of insulations as a function of density.

Strength. Specified compressive and flexural strengths also increase with specified density, as illustrated in Figures 3-4 and 3-5. Note that the compressive strength of polystyrene is called "compressive resistance" for foams and specified for a 10 percent deformation. Unlike many other materials, foams react to outside forces gradually. Under increasing compressive forces they become progressively thinner; there is no clear yield point at which they shatter. Since the application of compressive force must be stopped at some consistent point for comparison purposes, a deformation of 10 percent is selected as the standard. Fiberglass is omitted because its strength on either dimension is negligible.

Water vapor permeance. Water vapor permeance measures the rate at which vapor passes through the foam at a controlled vapor pressure. By convention, products with a permeance lower than 1 perm qualify as a "vapor retarder" (ASHRAE, 1993). Permeance this low is achieved by various plastic sheet products. Figure 3-6 graphs specified permeance for the foams as a function of specified density. The permeance of EPS is higher, and it changes more with density than does the perme-

Plastic Foams   73

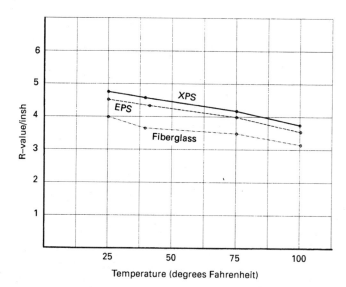

Note: EPS and XPS data points are for specified minimum R-values. Actual R-values may be higher.

Figure 3-3 Specified R-values of insulations as a function of temperature.

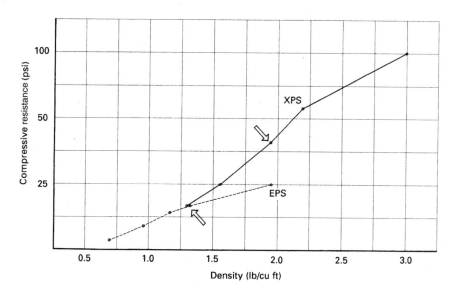

Note: ⇑ Arrow denotes most commonly used density.

Note: EPS and XPS data points are for specified minimum compressive resistance. Actual compressive resistance may be higher.

Figure 3-4 Specified compressive resistance of insulations as a function of density.

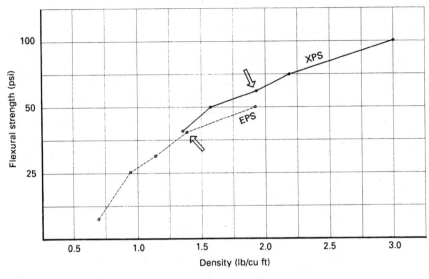

Note: ⇧ Arrow denotes most commonly used density.

Note: EPS and XPS data points are for specified minimum flexural strength. Actual flexural strength may be higher.

Figure 3-5 Specified flexural strength of polystyrene foam insulations as a function of density.

ance of XPS. The permeance of fiberglass is much higher than that of either of the polystyrene foams.

Water absorption. Water absorption measures the amount of water absorbed into the foam (as a percentage of total volume) when it is immersed in water for 24 hours. This is intended as a rough indicator of how much water the foam might absorb if wetted in the field. The number is generally high, however, since once installed, the foam will rarely, if ever, be immersed. The water absorption of EPS falls as the density increases (Figure 3-7). One can think of higher-density foam as fusing the beads more "tightly," thus resisting liquid penetration. For XPS, the relationship is flat. Both EPS and XPS resist moisture absorption, but the closed-cell structure and lack of voids of XPS between cells helps it resist moisture penetration somewhat more.

Water absorption is of interest partly because absorbed water reduces thermal resistance. Figure 3-8 plots the relationship. Some cite EPS' greater absorption and subsequent reduction in R-value as a reason for preferring XPS below grade. However, the reduction may be slight within the range of absorption that occurs in the field.

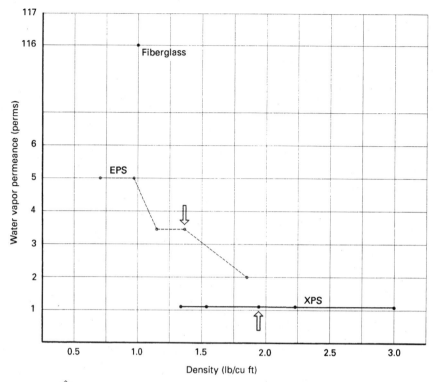

Note: ⇑ Arrow denotes most commonly used density.

Note: EPS and XPS data points are for specified maximum water vapor permeance. Actual permeances may be lower.

Figure 3-6 Specified water vapor permeance of insulations as a function of density.

Dimensional stability. The dimensional stability of the foams is measured by subjecting a sample to temperature cycles ranging between −40 and 158°F over a period of 7 days. According to the specification, it is not to vary in any linear dimension by a total of more than 2 percent. Most formulations of polystyrene foam expand and contract far less than the limit, as one would predict from their very low thermal coefficient of expansion (see Chapter 1).

Oxygen index. The oxygen index specification helps determine that the foam is adequately modified with a flame retardant. Technically, the oxygen index is the percentage of oxygen in the atmosphere that would have to be present for the given foam to sustain combustion. If the actual atmosphere were lower in oxygen than the foam's index, the foam would not burn unless constantly exposed to an outside fuel source. ASTM C578-95 sets the index minimum at 24 for all poly-

76 Components

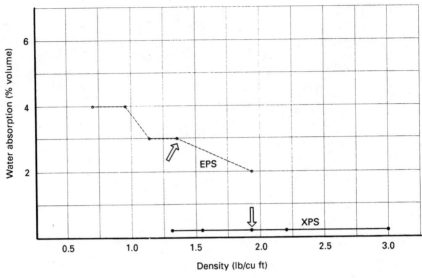

Figure 3-7 Specified water absorption of insulations as a function of density.

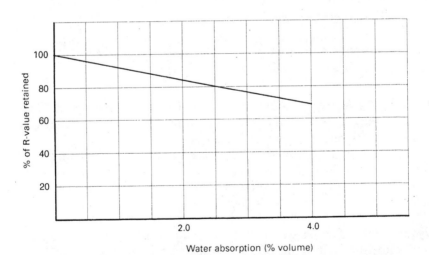

Figure 3-8 R-value retention of polystyrene foam insulation as a function of water absorption.

styrene foam types. Since the natural atmosphere includes less than 24 percent oxygen, foam with an index of 24 or more will not ordinarily sustain combustion.

Flame spread and smoke development. Although they are not included in the ASTM standard, building codes establish limits for insulation on two fire-related properties. These are the flame spread and the smoke development index (SDI). The flame spread measures the distance that a controlled flame carries across the surface of a material. Smoke development measures the amount of visible smoke the material emits in the same situation. The scale of each measure is calibrated by the performance of red oak, which is set equal to 100. The maxima allowed are 75 for flame spread and 450 for SDI. ICF manufacturers report values for their foams well within the limits of these requirements, as indicated in Table 3-2.

Note that the SDI will be higher the thicker the material. Here we report the SDI for a 2-inch thickness of foam, which is approximately the depth of a typical ICF face shell. Our data on flame spread and SDI also do not include any contribution from rigid ties or fastening surfaces.

Toxicity. EPS and XPS foam products are not classified or regulated as toxic materials by the EPA, OSHA, or FDA. Polystyrene foams are in fact approved for many applications that bring them in contact with food.

Polystyrene products do not contain any chemicals in amounts above the government agencies' established safety thresholds. Tests of molded EPS products have attempted to measure the presence of possible chemical fragments of polystyrene (styrene monomer and benzene) that might reside in the material in a free, unbound state. The amounts of these components turned out to be vanishingly small (Pesselman and Zhao, 1990).

Moreover, the components of polystyrene foam have proven to be stably bound within it. Tests on EPS cups have shown that the styrene monomer does not leach out when in contact with food. Scientists in a German study (Voss, 1987) measured the air of homes insulated with XPS to detect styrene residuals. They found none, even when they calibrated their instruments to make them five times more sensitive than originally designed.

Some of the blowing agents of the polystyrene foams may also reside in the final material. However, these also remain in small concentrations and are not considered or classified as toxic. As discussed in Chapter 4, when polystyrene foams are forced to burn, the by-products that result are considered no more toxic than those of wood.

Cost. The cost of the foams varies with many factors. Higher densities are generally more expensive. The point of purchase, order volume, and current market conditions are all influential. The prices in Table 3-2 apply to 1-inch sheet material purchased in small volume at retail. In practice, the more relevant cost is the price of actual ICF units. Although this price naturally includes the cost of materials, manufacturers set it based on many other factors as well. *The ultimate source of price information is therefore current quotations from ICF manufacturers.*

Recycling. Several of the ICF manufacturers advertise that much or all of the polystyrene that goes into their products is recycled. Contact the individual firm for information.

Polyurethanes

Different polyurethane formulations vary widely, and no specification for the material exists that is as widely accepted as ASTM C578 is for polystyrene foams. Therefore the description of individual polyurethanes generally consists of a full list of properties, as tested under contract or otherwise guaranteed by the manufacturer. However, the Society of the Plastics Industry (1994) has published recommended specifications for certain polyurethanes used in building insulation, which we reproduce in Table 3-4. While they apply to a different formulation of polyurethane from that used in ICFs, the specified properties are representative.

As apparent from Tables 3-2 and 3-4, polyurethane's strengths are in line with those of the polystyrenes. Flame spread and smoke developed are also similar and fall below the maxima prescribed by building codes.

TABLE 3-4 Specifications of Society of the Plastics Industry for Polyurethane Foam Used in Construction*

Property	Specified value
Density (lb/cu ft)	1.5–3.0
R-value per inch, aged, min	6.2
Compressive resistance, min (psi)	15
Flame spread, max	75
Smoke developed, max	450

SOURCE: Society of the Plastics Industry (1994).
*Actually written for spray polyurethane foam. ICFs use molded polyurethane foam, not spray. However, values of the properties are similar.

The R-value of polyurethanes is listed as "aged" in the table because it declines somewhat after manufacture. Exposed to air, polyurethane gradually leaks some of the entrapped blowing agent that provides it with its unusually high thermal resistance. Numerous tests suggest that unless the foam is faced with diffusion-retarding films, the R-values of polyurethanes decline over a period of 2 to 5 years and then stabilize (Society of the Plastics Industry, 1988). Initial R-values range from approximately 8 to 9 per inch. If the foam is faced with gas-impermeable diffusion barriers, R ultimately falls to approximately 7 or 8. If it is not faced, R stablilizes in the range of 5.6 to 6.2 per inch. Some manufacturers report aged R-values. These approximate the ultimate R-value of a foam by testing it after subjecting it to an accelerated aging procedure. Because of the large differences, it can be important to determine whether a given manufacturer is reporting the R of its foam product in an aged or an initial state and whether in a faced or unfaced condition. Although finishes on the ICF wall may retard the decline of the R-value to some degree, the generally recommended design value for polyurethane foam in wall insulation is 6 per inch (ASHRAE, 1993).

The polyurethanes tend to be more expensive than some other foams, as reflected in Table 3-2. This, however, does not necessarily translate into higher costs for ICF units made of polyurethane since some lower processing costs can offset the material cost.

The toxic potential of the polyurethanes used in ICFs is considered negligible. They do not employ formaldehydes (unlike some wood and fabric products used in construction). Similar to XPS, they encapsulate a blowing agent. But these agents are considered nontoxic and escape in only small quantities anyway. Some of the ICF manufacturers have their polyurethane formulations tested, and will provide reports verifying that toxins are not present in the material.

EPS-cement composites

The manufacturer-stated properties of these variants on EPS, used in two of the grid panel systems, are also listed in Table 3-2. They reflect the alleged greater density and strength and the somewhat lower R-value of these materials. The density figure is most relevant as a reflection of the greater weight of the final ICF units. It does not equate to other material properties as do the densities of the pure foams.

Test data on the other properties are not yet available. Concerning fire resistance, one of the manufacturers claims (ENER-GRID, 1995) that the material was virtually unaffected through a standard ASTM fire test lasting 2 hours (see Chapter 4). This great resistance to heat and flame would be consistent with the lower fire vulnerability of portland cement as compared with plastic. Unit costs of the material alone are also not available.

Given the low toxicity of EPS and cement separately, one would expect the composites to be similarly benign. The manufacturers employing composites claim that some or all of the foam beads used are recycled.

References

ASHRAE, 1993
    1993 ASHRAE Handbook of Fundamentals. New York: American Society of Heating, Refrigeration, and Air Conditioning Engineers.

ASTM, 1995
    "Standard Specification for Rigid, Cellular Polystyrene Thermal Insulation," ASTM Standard C578-92. In *1995 Annual Book of ASTM Standards.* Philadelphia, PA: American Society for Testing and Materials.

ENER-GRID, 1995
    ENER-GRID Technical Manual. Phoenix, AZ: ENER-GRID, Inc.

Langlais, Silberstein, and Sandberg, 1994
    Langlais, Catherine, Anne Silberstein, and Per Ingvar Sandberg, "Effects of Moisture on the Thermal Performance of Insulating Materials." In Heinz R. Trechsel, Ed., *Moisture Control in Buildings.* Philadelphia, PA: American Society for Testing and Materials. 1994.

Pesselman and Zhao, 1990
    Pesselman, Robert L., and Xunying Zhao, "Determination of Residual Styrene Oxide and Benzene from Polystyrene Products," Laboratory Project HLA 6001-477, Hazelton Laboratories America, Inc., Madison, WI, January 23.

Society of the Plastics Industry, 1988
    "Rigid Polyurethane and Polyisocyanurate Foams: An Assessment of Their Insulating Properties." In *Polyurethanes 88,* Proceedings of the Society of the Plastics Industry," 31st Annual Technical/Marketing Conference, Philadelphia, PA, October 18–21, pp. 323–337.

Society of the Plastics Industry, 1994
    "Spray Polyurethane Foam for Building Envelope Insulation and Air Seal," Publication No. PFCD/GS4-8/95, Polyurethane Foam Contractors Division, Society of the Plastics Industry, Inc.

Voss, 1987
    Voss, H., "Untersuchungen zur Styrol-Emission in mit Polystyrol-Hartschaumstoff wärmegedämmten Wohnräumen," *Kunststoffe,* Vol. 77.

Part

3

Design

The major tasks of designing with ICFs are the same as with other structural systems:

Establish requirements of the building

Specify layout, elevations, and finishes

Select components and materials

Estimate costs

Specify details of the structure to withstand the applied loads

Specify HVAC equipment to maintain the indoor environment

A single designer often executes all of these tasks. But the last two are sometimes performed separately by engineers, in coordination with the primary designer. Reflecting this division of responsibilities, we cover the early design tasks here in Part 3 and structural and HVAC engineering in Part 4.

The chapters in Part 3, like the rest of the book, present only generalizations on typical ICF design and construction practice. Each ICF system is proprietary and unique, and each manufacturer has unique specifications for proper use of its system. Moreover, these specifications can change over time. **The specifications and directions of the ICF system manufacturer always take precedence over the information presented in this book.**

Chapter

4

Detailing Considerations

Certain harmful natural forces can be counteracted with appropriate design. Although a detailed understanding of them and how they affect ICF walls is not necessarily essential, it is useful for those who wish to hone their design skills, or who are interested in broadening their related knowledge.

Fire

The course a building fire follows and the harm it causes are complex phenomena influenced by a large number of factors. While the make-up of exterior walls is only one such factor, minimizing its potential contribution is desirable.

There are four major concerns about the performance of exterior wall systems in a fire. The first is that they will fail structurally. The resulting collapse could injure occupants, increase property damage, and drop additional fuel (roof, contents of upper floors, etc.) into the fire and thereby intensify it before fire control measures take effect.

The second is that the walls will allow fire originating on one side of the wall to pass through to the other. This is of concern when the fire originates outdoors, as does a brush fire, or a fire in one part of a multiunit building. Since ill consequences are limited if the fire does not spread indoors or to adjacent dwelling units, a wall system that prevents passage until help arrives is preferable. This is the reason why fire-resistant walls are required between dwelling units in multi-family buildings.

The third concern is that materials in the wall will burn, adding fuel and aggravating the fire and its resulting damage.

The fourth is that the materials, when subjected to fire, will emit fumes that asphyxiate occupants or incapacitate them before flames reach them. Some experts do not consider this risk to be extremely important (Andrews, 1992). Their reasoning is that any occupants slow to exit will likely succumb to fumes from the building's contents (which usually burn first) rather than from the walls.

Structural failure and fire passage

A physical test addresses the first two concerns. ASTM E119 (ASTM, 1995), commonly known as the "fire wall test," is actually used most often to determine a wall's suitability as a partition between adjoining living units. Most codes require it only of walls that will be used for that purpose. However, it provides information useful to assessing both a wall's structural integrity in fire and its resistance to allowing fire to pass.

In the test, laboratory personnel subject one side of a sample wall to a controlled gas flame. They record the elapsed time as soon as either (1) the opposite side of the wall exceeds a specified temperature (one likely to cause spontaneous combustion in common household materials), (2) a sample of organic material (cotton waste) placed along the opposite side ignites, or (3) the wall fails sufficiently to allow flames or smoke to pass through.

Sample wood frame walls with 2×4 studs every 16 inches on center, conventional finishes, and fiberglass insulation receive a fire resistance rating in ASTM E119 tests of 1 hour (Underwriters Laboratories Inc., 1988). Usually they collapse or burn through soon after this much time has elapsed. Manufacturers of at least 10 of the ICFs have had their systems tested, and all received ratings of 2 to 4 hours. In none of the ICF tests did the wall fail structurally. In fact, some systems tested for 2 hours could have continued (they had violated none of the three test conditions), but the sponsors had paid for only that amount of time. Note that all of the tested walls were either flat systems, uninterrupted grid systems, or one of the interrupted grid systems with units made of a foam-cement composite. There are no reported fire test results for interrupted grid blocks made of a pure foam or any post-and-beam systems.

Potential fuel source

The potentially flammable material in ICF walls is the foam. As discussed in Chapter 2, virtually all of the foams used in ICFs are manufactured with a flame-retardant additive. Tests of these foams indicate that, unlike lumber, they will not support combustion. That is, if lit they will not by themselves burn. If they are subjected to constant flame from an outside source (e.g., burning gas or wood), they can burn.

The Steiner Tunnel flame spread test (ASTM E84) is a widely accepted method of judging the tendency of flame-exposed materials to burn. As discussed in Chapter 3, the foams used in ICFs test at a much lower flame spread than common wood.

Combustion emissions

If the foams used in ICFs are forced to burn, their combustion products are judged to be no more toxic than those of wood (Grand, Kaplan, and Hartzell, 1986). It is difficult to be precise on this topic, however. The combustion products of any particular material vary with conditions, and the toxicity of any particular combustion product is still the subject of debate. In complete combustion the foams produce two products considered potentially toxic: carbon dioxide and carbon monoxide. However, they yield concentrations that are about one-tenth those from wood and wood products. Wood, moreover, produces a variety of others, such as aldehydes and ketones.

In incomplete combustion both materials also produce carbon smoke, with the foams producing more. The smoke obstructs vision and, like the other products mentioned, probably has some toxic effects on humans when inhaled in large doses. The Steiner Tunnel test includes a procedure to measure the visible smoke from the burning subject material with a photoelectric eye. This is quantified as the SDI. As discussed in Chapter 3, some ICF foams can produce more smoke than wood, but all are below the code-allowed maximum level for insulation products.

Design implications

The first apparent implication of the combustion data for design is that ICF walls have advantages in fire compared with wood frame. They should retain their structural integrity longer, prevent fire movement across walls longer, and spread flame less.

The second is the importance of adequately covering the foam. Recommendations here are the same as they are for frame. Combustion of the foam can create smoke harmful to occupants. To delay this, sheath indoors with gypsum wallboard or equivalent. The delay gives the occupants time to exit safely. Virtually all code jurisdictions require at least a 15-minute thermal barrier over foam inside a living space, and some require it in basements and garages as well. In areas with high risk of outdoor fires, covering the exterior with a fire-resistant finish is also advisable. Effective popular alternatives are stucco and masonry veneer, with a minimum of exposed wood such as for trim or eaves (Underwood, 1995).

Wind

High winds, especially from hurricanes and tornadoes, can cause failure in structural members of houses. In addition to the direct property damage that results, flying debris or structural collapse can injure occupants.

After hurricanes Andrew (South Florida) and Iniki (Hawaii) of 1992, two study teams conducted surveys of the affected areas to determine which design and construction practices are most susceptible to wind damage. A group headed by Dr. Ronald Zollo of the University of Miami visited the areas ravaged by Andrew (Zollo, 1993), and the research arm of the National Association of Home Builders surveyed both (NAHB Research Center, 1993).

Both studies reached similar conclusions. The most common damage was the loss of all or parts of the roof in approximately 25 percent of the homes in the direct hurricane path. This loss could almost always be traced to the wind acting directly on the roof, either under the eaves or against a gable end, or to windows blowing in and the wind entering through them to push up on the roof from inside. After loss of the roof diaphragm, houses with concrete walls rarely suffered any significant further damage. In contrast, many roof-damaged frame houses experienced severe wall damage (although the exact proportion experiencing damage was not reported). The explanation given for this result is that the greater inherent weight and strength of reinforced concrete walls are effective in resisting the wind's high, sustained uplift and transverse forces.

Components are available to reinforce the connections of frame houses and make them more resistant to wind damage. However, they add cost to the home, as discussed in Chapters 1 and 8. In addition, the University of Miami report claims that wood frame construction is less "forgiving" than concrete, in the sense that variations in material and workmanship are more likely to compromise the resistance to wind damage.

A few design implications follow from this research. First, hip roofs are more wind-resistant than gable or gambrel roofs. That appears to apply with any wall system. Second, holding down the roofing members with steel straps is more wind-resistant than nailing them to the wall. Chapter 6 presents details for doing this with ICFs. It happens to be particularly economical with a concrete wall. And third, concrete walls resist wind better than conventional frame. That benefit is automatic with an ICF, although Chapters 9 and 10 discuss measures that can be used to raise the structure's wind resistance to even higher levels. Fourth, stout shutters or high-strength windows increase survival of the windows themselves and retard roof damage. This also holds regardless of the wall system.

Moisture

The major ways in which exterior walls can contribute to moisture problems in small buildings are by (1) providing a conducive environment for the growth of mold and mildew inside the wall assembly or on its interior surface, (2) allowing interior humidity to rise to, and persist at, a high level, (3) allowing moisture entering the wall to damage it, detracting from its aesthetics or structural soundness, and (4) allowing moisture to pass through the wall and damage other, interior parts of the building.

As we discuss next, the first three classes of problem are unlikely in ICF houses due to the physical properties of the wall assembly. The fourth is also not common in ICFs, but it apparently can occur around certain types of opening details that have caused problems on frame as well. Some appropriate details are recommended.

Mold and mildew

Mold and mildew are fungi that grow in a moist atmosphere. They tend to form on the interior side of exterior walls in northern (heating) climates when that surface has cold spots. Interior air contacting relatively cool points on the surface becomes colder, and hence increases in relative humidity. If severe, moisture in the air condenses to form beads of liquid water on the wall. On frame walls these things tend to occur along the locations of the studs. The studs act as a *thermal bridge:* because of their relatively low R-values the studs permit heat to leak out rapidly, cooling themselves and the interior wall surface they contact (Figure 4-1).

Conversely, in warmer (cooling) climates mold and mildew often form when moist exterior air infiltrates the wall and reaches the materials in contact with the interior (notably the wallboard). These ma-

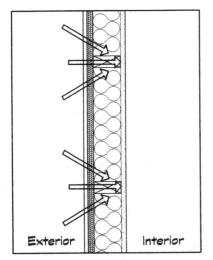 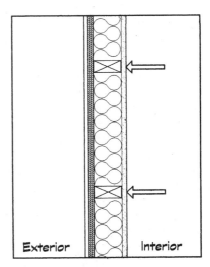

Figure 4-1 Warm air from outside condensing on cool wallboard (left). Warm air from inside condensing along thermal break of studs (right).

terials are usually cool as a result of air conditioning. Relative humidity rises, and the fungi grow on the exterior surface of the wallboard or adjacent lumber.

Once these plants form, they can discolor the materials beneath them, deteriorate the materials' structures, and emit particles that cause unpleasant odors or affect the health of building occupants.

An ICF wall is unlikely to permit significant mold or mildew growth on either side of the interior wallboard. There are few thermal breaks in the wall. This might be of some greater concern with systems that employ steel ties, but their R-values suggest that the thermal bridging effect is not extremely large.

As for air-conditioned houses in warm climates, the low infiltration rates of ICF walls should prevent humid air from reaching the interior wallboard. As discussed in Chapter 3, the concrete and foams used are good water vapor diffusion retarders as well as air barriers.

Interior humidity

In humid climates, infiltration of moist air from outdoors can keep interior humidity at levels that are uncomfortable or unhealthy to occupants, favor mold and mildew growth throughout the indoors (not just at cold spots), and eventually lead to damage to water-sensitive contents of the building. As already noted, ICF walls are relatively impermeable to both air and water vapor infiltration. Although it may be possible for any house to experience high levels of humidity from infiltration, building with ICFs should reduce any contribution of the walls to the problem because of their low infiltrations. The difficulty is more likely to occur as a result of poor design or installation of the roof or doors and windows.

Damage to wall or interior

The exterior wall system may contribute to damage if it allows water to pass. With frame, entering water can damage the wall assembly itself. If it reaches the interior, water can also damage other structural members and contents that are prone to rotting or dissolving. It could also contribute to the type of mold and interior humidity problems described previously.

The first aspect of this problem—damage to the wall itself—appears unlikely in ICFs. As discussed in Chapter 1 under "Durability," the important properties of the materials in an ICF wall (notably R-value and structural strength) are little affected by exposure to moisture. This contrasts with the wood of a frame wall.

Nor do we have confirmed reports of water passing through ICF walls themselves to the interior. In principle it might be possible for water contacting the exterior foam (as the result of some failure in the

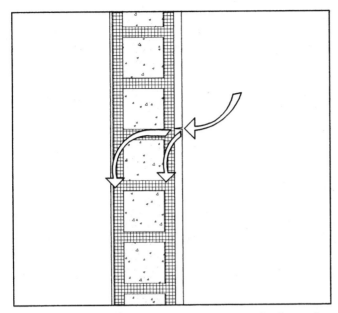

Figure 4-2 Alternate paths of moisture entering the foam of an ICF wall through a gap in the exterior finish.

finish) to move sideways through capillary action, rather than downward because of gravity. Figure 4-2 shows the alternatives graphically. But independent testing of molded EPS, the predominant foam used in ICFs, indicates that water moves quickly downward through the material (Kenney, 1996). Moreover, those involved in the testing argue that a concrete layer will impede the water from moving further sideways if it gets that far. In contrast, some experts argue that water has moved sideways through the foam sheet of exterior insulation and finish system (EIFS) installations (Lstiburek and Carmody, 1993). If this is correct, in systems that have continuous foam paths between outside and inside (specifically, systems of interrupted grids or post and beams), water penetrating the exterior finish might eventually arrive indoors. It is therefore advisable to practice the same sort of care with exterior finish that is necessary over frame.

Similarly, there are some unconfirmed reports of groundwater passing through basement walls by traveling along the joint between an ICF's rigid tie and the concrete. This reconfirms the importance of treating ICF walls below grade with the same measures used on conventional foundations (see Chapter 6 for details).

There have been a few reported instances of water penetrating ICF walls at the window joints. These appear to result from the same causes that have recently led to significant damage in numerous wood frame homes (Energy Design Update, 1995). In frame, the problem

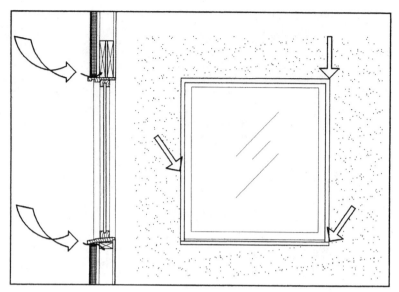

Figure 4-3 Rainwater entering the cavity of inadequately flashed or sealed window joints.

houses employed an EIFS. An EIFS consists of a thin-coat stucco applied over a layer of foam sheathing. The most recent spate of problem reports came from North Carolina, but some have also originated in other areas with a wet climate and a building code that requires a vapor retarder on the inside of a frame wall (Figure 4-3). Apparently, rainwater enters the wall around openings or joints where flashing or sealing were incorrect or inadequate. This is particularly a hazard with an EIFS or other stucco finish, for two reasons. Window joints are often protected with sealant alone, and the nature of the system can require some extra effort to extend flashing a few inches beyond either side of the opening, as it should. Once inside, the water eventually rots or dissolves the sheathing (depending on whether it is a wood or gypsum product), or rots the studs.

A few ICF houses have similarly been reported to allow water indoors around windows. The ICF walls themselves showed no signs of damage. Otherwise the circumstances were nearly the same as with the frame houses (Crandell, 1996). The ICF homes employed a thin-coat stucco exterior finish. Notably, the failures occurred only in regions and walls subject to significant driven rain (blown sideways by the wind). The conclusion was that the cause was sealant failure or flashing not extending beyond the sides of the openings, on a window detail without backup joint protection.

Design implications

The general high resistance of ICF walls to condensation, moisture penetration, and moisture damage leaves little need for special detail work to protect from these. In two areas precautions are wise: exterior finish and window connection on homes finished with stucco.

Although we have no reports of water penetration of ICF walls themselves, the theoretical possibility that it could occur through capillary effects favors careful construction of the exterior finish. Put simply, the foam should not get wet any more than a frame wall behind its exterior finish should. Thus the level of attention on related details should match what is practiced on frame.

Conservative detailing should also minimize the probability of water penetrating the window joints of houses finished with stucco. Some recommended details are given in Chapter 6.

Insects

We have not yet documented an occurrence of insect infestation in an ICF home. There are also some reasons to think it would be unlikely. However, infestations have appeared in other foam products used in building construction. This warrants caution in projects believed to be at high risk. The following section examines the major insect threats to provide an understanding of how and why the preventive measures (detailed in Chapter 6) should work.

The insects that cause the most damage to wood homes in the United States are the subterranean termite and carpenter ant, in that order. The nature of the threat they pose and the measures to fend them off are different.

Subterranean termites

Although the foams used in ICFs provide no nutrition to termites, there is a concern that they might use them as a pathway to reach and attack the wooden members of a house.

Subterranean termites live in colonies that scavenge for wood, their primary food. As a group they seek wood deposits and establish paths by which they can routinely travel to them from their nests in the ground. Although they have appeared in virtually every corner of the United States and Canada, they are more common in warmer climates.

Subterranean termites are attracted to the lumber of conventional frame houses. To reach the wooden structure they build mud tunnels or "tubes" up along the concrete foundation. These form paths through which their colony can routinely travel to the food source without exposure to attack. Once they have established routes into a house, the termites' effects range from annoying noise in the walls to (eventually) serious structural damage of the wood members.

Regular termite inspection by pest control operators (PCOs) is routine in much of the South. Although there are various methods of determining whether the insects are entering a house, when the structure has a stem wall or basement foundation, one of the simplest, most reliable and inexpensive is to look for mud tubes. Once signs of termite entry are found, or as a preventive measure, a common treatment consists of putting insect-killing chemicals into the soil around the foundation (as discussed in Chapter 1). The termites cannot reach the house without contacting the pesticide.

Recently PCOs have issued warnings about foundations wrapped in foam. These warnings, by extension, are the source of concern about insects and ICFs. Two surveys of PCOs revealed that a high percentage of them had, at least once, encountered houses with foam on the foundations that had been visited by termites (Mar-Quest Research, 1993; Smith and Zungoli, 1995). Many of the PCOs also reported that they believed the appearance of termites to be linked to the use of foam.

The data do not include estimates of how frequently homes with foam on their foundations become infested, or whether they do so with greater or lower frequency than homes without foam cladding. Moreover, the data apparently covered primarily conventional foundations sided after their erection with foam sheet. But underlying the concerns are potential problems that could perhaps occur in ICF foundations as well.

PCOs and scientists theorize that foam cladding on the exterior of a foundation wall provides termites with a path to the wood in a house that avoids both inspection and conventional soil treatments (as depicted in Figure 1-10). The thinking goes that termites can enter the foam below the treated soil and tunnel through it up to the wood of floor decks and roofs. Since they are below the surface, no tubes would be visible. The result is that they would be more likely to go undetected and are less subject to common treatments once they are detected.

However, ICFs have some attributes that may discourage termites. Below and at grade they are usually covered with dampproofing or waterproofing. Termites may not eat through these materials. If the above-grade walls are also of ICFs, the insects generally have to travel farther to get to wood. Assuming the bucks are made of pressure-treated lumber (which termites rarely attack) and there is no foam pathway (e.g., a foam web) to the interior, they may need to climb to the roof. In addition, preliminary results from ongoing research indicate that termites tend to climb in the gap between concrete and foam rather than eat long tunnels through it (Hansen, 1996; Williams, 1996). But most ICF walls have little or no gap: the concrete adheres to the foam. So the need to tunnel the entire distance might discourage the insects.

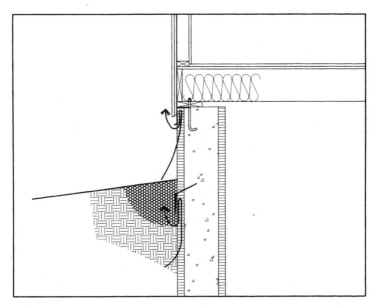

Figure 4-4 Blocking subterranean termite path through foam with termite shields.

Nonetheless, users in regions of high termite risk may prefer to take extra precautions. Chapter 6 includes details for buildings subject to high termite threat. These generally involve some form of break in the foam that forces the insects to leave it. Figure 4-4 depicts the operation of such a break. One may place the break just below grade, a little above, or both. It blocks the termites' path through the foam, forcing them to exit so that they must enter the treated soil (below grade) or build mud tubes that inspectors can detect (above). As discussed in Chapter 6, the North Carolina State Building Code now requires a particular type of break on new construction.

Also of interest are foams treated with pesticides. Some foam manufacturers have come up with additives designed to kill insects that tunnel through the material. The pesticides are available in some ICFs, and are currently undergoing tests to determine their effectiveness. Over time the number of pest-treated systems may grow.

Carpenter ants

More prevalent in northern climates is the carpenter ant (Akre and Hansen, 1990). Unlike termites, the carpenter ant does not feed on wood. Colonies of the insect dig channels in wood simply to nest there. It has also been found nesting in foam insulation in some houses, eventually hollowing out large sections (Andrews, 1992). However, there are no data to determine whether they attack structures with foam insulation any more or less often than they do wood frame buildings. In addition, we have to date uncovered no instances of them attacking the foam of an ICF house. Were they to do so, the potential damage includes loss of insulation, loss of backing for the finish materials, and loss of any adjacent wooden members. The concrete walls, however, should remain structurally intact.

Since carpenter ants travel above ground, experts believe they can be prevented, detected, and treated in ICF houses the same ways they are in frame houses (Hansen, 1996). Chapter 6 provides details for buildings deemed at high risk of attack.

References

Akre and Hansen, 1990
    Akre, R. D., and L. D. Hansen, "Management of Carpenter Ants." In Robert K. Vander Meer, Klaus Jaffe, and Aragua Cedeno, Eds., *Applied Myrmecology: A World Perspective*. Boulder, CO: Westview Press, pp. 693–699.

Andrews, 1992
    Andrews, Steve, *Foam-Core Panels and Building Systems*. Arlington, MA: Cutter Information Corp.

ASTM, 1995
    "Standard Test Methods for Fire Tests of Building Construction and Materials," ASTM Standard E119-95a. In *1995 Annual Book of ASTM Standards*. Philadelphia, PA: American Society for Testing and Materials.

Crandell, 1996
    Crandell, Jay H., "Investigation of Moisture Damage in Insulating Concrete Form Homes Sided with Exterior Insulation Finish Systems in Eugene, Oregon," Report to the Portland Cement Association from the NAHB Research Center, Upper Marlboro, MD, May 1.

Energy Design Update, 1995
    "Severe Rotting Found in Homes with Exterior Insulation Systems," Vol. 15, No. 12 (December), pp. 1–3.

Grand, Kaplan, and Hartzell, 1986
    Grand, A. F., Harold L. Kaplan, and Gordon E. Hartzell, "A Literature Review of the Combustion Toxicity of Expanded Polystyrene," Final Report, SwRI Project No. 01-8818-507, Southwest Research Institute, San Antonio, TX, May.

Hansen, 1996
    Hansen, Laurel D., Biology Department, Spokane Falls Community College, Spokane, WA. Personal Communication.

Kenney, 1996
    Kenney, Russel J., R. J. Kenney Associates, Inc., Plainville, MA. Personal communication.

Lstiburek and Carmody, 1993
    Lstiburek, Joseph W., and John Carmody, *Moisture Control Handbook*. New York: Van Nostrand Reinhold.

Mar-Quest Research, 1993
    "Quantitative Analysis of Foundation Insulation's Impact on Subterranean Termite Control," Mar-Quest Research, Inc., Unpublished Report of Project No. 260-3, prepared for DowElanco, August.

NAHB Research Center, 1993
    "Assessment of Damage to Single-Family Homes Caused by Hurricanes Andrew and Iniki," Report prepared for U.S. Department of Housing and Urban Development and Office of Policy Development and Research, Contract HC-5911, September.

Smith and Zungoli, 1995
    Smith, B. C., and P. A. Zungoli, "Rigid Board Insulation in South Carolina: Its Impact on Damage, Inspection and Control of Termites," *Florida Entomologist,* Vol. 78, No. 3 (September), pp. 507–515.

Underwood, 1995
    Underwood, John, "Fire-Resistant Details," *Fine Homebuilding,* July 1995, pp. 90–94.

Underwriters Laboratories Inc., 1988
    Fire Resistance Directory. Northbrook, IL: Underwriters Laboratory.

Williams, 1996
    Williams, Lonnie, Ridge Mountain Wood Protection Services, Gulfport, MS. Personal communication.

Zollo, 1993
    Zollo, Ronald, "Hurricane Andrew: August 24, 1992, Structural Performance of Buildings in Dade County, Florida," Technical Report No. CEN 93-1, University of Miami, Coral Gables, FL.

Chapter

5

System Selection

The available ICF systems have differences that can be important in specific building projects. Table 5-1 includes information on items commonly considered in selecting a system. However, some very important characteristics of ICF systems change so rapidly from time to time and place to place that we cannot quantify them precisely. They are therefore not given in Table 5-1. Therefore, since the systems are changing, the reader should always consult manufacturers to verify product data.

This chapter discusses the various selection criteria. For most of them no one system or type of system is clearly superior. There are advantages and disadvantages to each. Therefore we provide a number of tables listing the scopes, advantages, and disadvantages of the alternatives.

Important Criteria Not Quantified

Foremost among the items that change too rapidly to quantify are:

Geographic availability

Service level

R-value

Dimension consistency

Price

Given the importance of these to system selection, a good approach is to choose a few attractive systems based on the criteria that are quantified (discussed later in this chapter) and then call each manufacturer to get specifics on the criteria not yet quantified.

TABLE 5-1 Commonly Considered Criteria for System Selection

System	Concrete content	Unit area	Foam webbing	Integral fastening surface	Foam class	Specialty units	Unit assembly	Research reports
Flat panel systems:								
Lite-Form (preassembled)	High	Large	None	Rectangles	Optional	None	Pre-	None
R-FORMS	Med–high	Large	None	Rectangles	Optional	None	User	None
Grid panel systems:								
ENER-GRID	Low	Med–large	Composite	None	Composite	Some	Pre-	None
RASTRA	Low	Med–large	Composite	None	Composite	Some	Pre-	ICBO
Post-and-beam panel systems:								
Amhome	Low	Large	Wide	Vertical	EPS	None	User	None
ThermoFormed	Low–med	Large	Narrow	Horizontal	Optional	None	Pre-	None
Flat plank systems:								
Diamond Snap-Form	Med–high	Medium	None	Rectangles	EPS	Some	User	None
Lite-Form (nonassembled)	High	Medium	None	Rectangles	Optional	None	User	None
Polycrete	Med–high	Medium	None	Horizontal	EPS	None	User	CCMC
Quad-Lock	Med–high	Medium	None	Rectangles	EPS	None	User	ICBO
Flat block systems:								
Blue Maxx	Med–high	Medium	None	Vertical	EPS	Many	Pre-	BOCA, ICBO, SBCCI, CCMC
Fold-Form	Med–high	Medium	None	Vertical	EPS	None	Pre-	None
GREEN-BLOCK	Medium	Medium	None	Vertical	EPS	Many	Pre-	ICBO, CCMC

System							
SmartBlock VWF	Med–high	Small	None	Vertical	EPS	Some	User	BOCA, ICBO, SBCCI
Grid block systems:							
I.C.E. Block	Medium	Medium	None	Vertical	EPS	Some	Pre-	BOCA, ICBO, SBCCI, CCMC
Insulform	Medium	Medium	Narrow	Horizontal*	EPS	Some	Pre-	None
Modu-Lock	Medium	Small	Narrow	None	EPS	Some	Pre-	None
Polysteel	Medium	Medium	None	Vertical	EPS	Some	Pre-	BOCA, ICBO, SBCCI, CCMC
Reddi-Form	Medium	Medium	Narrow	Horizontal*	EPS	Some	Pre-	BOCA, ICBO
REWARD	Medium	Medium	None	Vertical	EPS	Some	Pre-	None
SmartBlock SF 10	Medium	Small	Narrow	None	EPS	Some	Pre-	BOCA, ICBO, SBCCI
Therm-O-Wall	Medium	Medium	None	Vertical	EPS	Some	Pre-	None
Post-and-beam block systems:							
ENERGY LOCK	Low	Small	Wide-narrow	Vertical*	Polyurethane	Some	Pre-	None
Featherlite	Low–med	Very small	Wide-narrow	None	Polyurethane	Some	Pre-	ICBO
KEEVA	Low–med	Medium	Wide-narrow	None	EPS	Some	Pre-	ICBO

*Optional.

Availability

Geographic availability changes as the expanding ICF companies line up new production facilities. In principle any system can be shipped to job sites in any location. But transportation costs and delivery times mount with distance. The majority of projects employ a system manufactured within 500 miles. In some cases the prices of products sold by distant companies are low enough that they are attractive despite high shipping costs.

Service

The level of service from the manufacturer also varies. This is especially important for those who have not built extensively with ICFs before. The forms of service cited as most helpful are: site visits by field representatives, technical manuals, technical assistance by telephone, and code documentation. The availability of these from a manufacturer changes over time. The only reliable method of determining current local service levels is to speak directly with the company and its recent customers in your area.

R-values

Although actual R-values do not change frequently, methods of determining them do. There are several methods of estimating the thermal resistance of a wall system (see Chapter 11), with the result that different manufacturers publish R-values that are not strictly comparable. They may also change the R-values they report as results of newer calculations or tests become available. The most reliable recourse is to question manufacturers directly about a reported R-value and its method of determination. See Chapter 11 for a detailed discussion of alternative calculations of R-value and their interpretation. All current ICF systems are significantly more energy-efficient than even 2×6 frame construction.

Dimension consistency

The dimensional tolerances of ICFs are generally tight enough that they do not influence construction. However, contractors have reported that some batches of units had enough dimension variation to slow work. They also suggest that this has been less frequent with some brands than others, although it appears to be on the decline in general. The best source of information on this subject is recent local users of your candidate systems.

Price

Prices can also change over time and differ with location. Most quotes contractors reported for the price they paid for ICF formwork were in the range of $2.00 to $3.00 per square foot, although we received some lower and some higher ones. The exact price you pay will vary with such factors as the manufacturer's current policy, transportation cost, and order size. Obtaining a direct quote is therefore advisable.

A related consideration is that higher-priced systems often have extra or unusual features that can be valuable, such as a wide range of specialty units, special foams, or a high degree of technical documentation. These enter into several of the criteria discussed later in this chapter.

Concrete Content

The amount of concrete in a square foot of wall varies widely. As a group, flat systems are most concrete-dense, followed by grid systems. Several systems may have different concrete contents per square foot, depending on which unit is used. Wider units contain more concrete. Post-and-beam block systems can have either a medium (comparable to grids) or a low concrete content, depending on how many of the cavities one chooses to fill.

Table 5-2 summarizes the major pluses and minuses of different concrete contents. Low concrete content saves some money on concrete and the labor and equipment to place it. At $50 per cubic yard, concrete costs would be approximately $0.25 to $0.50, $0.50 to $1.00, and $1.00 to $1.75 per square foot of wall for low-, medium-, and high-concrete-content systems. Associated labor and equipment costs are less easily calculated, and probably vary less since they have fixed components.

TABLE 5-2 Concrete Content Alternatives

Concrete content (cu yd/sq ft)	Limitations	Advantages	Disadvantages
Low (0.005–0.01)	Above-grade, normal structural requirements	Lower concrete cost, faster concrete placement	Lower thermal mass
Medium (0.01–0.02)	None	Moderate thermal mass	
High (0.02–0.35)	None	High thermal mass	Higher concrete cost, slower concrete placement

TABLE 5-3 Unit Area Alternatives

Unit area (sq ft)	Limitations	Advantages	Disadvantages
Very small (0.8–1)	None	Less cutting length, easily handled	Cutting more units, more stacking motion, more glue or tape (if used)
Small (1–3.5)	None		
Large (4–10)	None		
Very large (10–32)	None	Cutting fewer units, less stacking motion	More cutting length, more cumbersome

Higher concrete contents have proportionately more thermal mass (see Chapter 1). They can also achieve higher structural strengths. However, all but the very lowest-content configurations can meet common residential requirements, including basement construction and high-wind and seismic loadings. For precise concrete contents of each system see Table 8-3.

Unit Size

Table 5-3 lists reasons for preferring systems with different size standard units. The unit size, either large or small, does not limit the range of possible applications of an ICF. Units with a smaller face shell area can require less total lineal feet of cutting, since they can more often be stacked right up to openings and corners. The chances of needing a cut in a system with large units are higher. Conversely, the total number of units that must be cut will be higher in small-unit systems. Smaller units are more easily handled, but workers must handle a larger total number of them. If glue or tape are used to reinforce interconnections, small units will need more of them.

Foam Webbing

Foam webs create a foam path through the concrete in a finished ICF wall. The size of this path varies from system to system. As noted in Table 5-4, systems having no foam webs create a continuous layer of concrete at the center of the wall. Some view this as a barrier to potential penetration by moisture, pests, fire, and physical impact. However, to date such problems have not been reported frequently for any ICF systems (see Chapter 1).

Conversely, foam webbing allows easy cutting through the wall after

TABLE 5-4 Foam Webbing Alternatives

Foam webbing	Limitations	Advantages	Disadvantages
None	None	Continuous concrete layer	Difficulty of postpour penetrations
Narrow	None	Ease of small postpour penetrations	Small concrete gaps, difficulty of large postpour penetrations
Wide	Above-grade normal structural requirements	Ease of all postpour penetrations	Large concrete gaps
Composite	None	Ease of small postpour penetrations	

the concrete is poured. This can be useful for running plumbing and electrical lines between indoors and out. Up to 2-inch-diameter cable or pipe can go through narrow foam webbing, and any common diameter through wide webs. In systems without foam webbing, crews usually provide for penetrations by installing sleeves of PVC pipe before the pour, or they drill the hardened concrete subsequently. In practice this does not appear to be costly or logistically difficult.

The limitations of wide-web systems are similar to those of systems with low concrete content. Indeed, the systems tend to be the same ones. They are not generally appropriate for below-grade use, or for rare situations requiring exceptionally high strength. Note, however, that all of the post-and-beam block systems will be narrow-web if all of their cavities are filled.

Systems with units made of foam-cement composites (ENER-GRID and RASTRA) are classified separately. Their webs can be cut postpour with ordinary wood tools to provide for penetrations, but with somewhat more effort than ordinary foam. The webs allow for cutting penetrations of a little over 4 inches in diameter. On the other hand, the resistance of the material to the unwanted penetrations of pests, fire, and impact should be somewhat better than that of the pure foams.

Fastening Surfaces

Whether an ICF has fastening surfaces, and what type, does not limit its range of applications (Table 5-5). As noted in Table 2-1, some ICFs have none. Others provide rectangular surfaces arranged in a grid, every 8 or 12 inches on center vertically and horizontally. The remainder have strips of steel, solid plastic, or wood that run at regular intervals (5 to 48 inches), either vertically or horizontally.

TABLE 5-5 Alternative Fastening Surface Status

Fastening surface	Limitations	Advantages	Disadvantages
None	None	Ease of cutting	Difficulty of wall connections
Rectangles	None	Ease of certain connections	Cutting effort
Vertical	None	Ease of locating horizontal attachments, similarity to stud placement	Cutting effort
Horizontal	None	Ease of locating vertical attachments	Cutting effort

Units having no fastening surfaces are easy to cut. Sometimes cutting units with rigid ties is a little more awkward. Cutting steel ties may require a special tool or blade. Note also that manufacturers of some systems (both with and without fastening surfaces) sell steel or plastic plates or strips that one can attach to the formwork (they lock into the concrete) to create a fastening surface where desired.

The major advantage of built-in fastening surfaces is that they provide convenient, reliably positioned locations for attaching such necessary items as temporary bracing, siding and other exterior-finish components, wallboard, and interior walls and trim.

Like the studs of a frame wall, fastening surfaces arranged as vertical strips have the advantage of allowing items that will be oriented horizontally to go at any height. Examples include clapboard siding, most vinyl sidings, and horizontal pieces of trim. Horizontal surfaces allow similar freedom in placing vertically oriented items, such as vertical trim and interior wall ends. Where items to be attached do not overlap a fastening surface, one can use connection arrangements like those used on systems without fastening surfaces. One can forecast the most useful geometry of fastening surfaces for a particular project from the wall attachments planned. Table 2-1 provides more exact data on fastening surface locations and materials.

Foam Class

As detailed in Chapter 3, different systems make their units of different plastic foams. Most use EPS, some allow the use of any sheet foam (EPS or XPS), two use polyurethane, and another two use a foam-cement composite.

Some designers and builders express a preference for one or another of these based on their properties, discussed in detail in Chapter 3. Table 3-1 lists the foam employed in each system to assist those for whom the material is a relevant selection criterion.

However, it is usually the properties of the final unit and the completed wall it creates, rather than those of the foam, that are most important. For example, the R-value of a unit will depend on the thickness of the foam as well as the type. And the total wall energy efficiency will also depend on thermal mass. Similarly, the cost of the foam used is not a good predictor of the cost of a system's units, and the cost of the final wall will depend on the ease of assembly as well as the unit cost.

Some systems can use any sheet material. With these, the decision of which system and which foam to use can be divorced somewhat. In these cases, foam selection is a more direct comparison of the costs and properties of the alternative materials.

Specialty Units

While some systems have no specialty units, most offer at least a 90° corner. A few systems offer several specialty units for use in multiple situations. However, not having special-purpose units does not limit a system's range of applications (Table 5-6).

Using no specialty units reduces some confusion, ordering, waste, and inventory costs. Systems offering no specialty units are sometimes less expensive.

Conversely, specialty units in effect perform certain operations in the factory instead of the field. In this way they reduce the amount of field labor required (and the associated construction time and cost). They also should reduce the variability of workmanship, since the units are produced in standard fashion on a machine. Note also that one may choose a system with specialty units without choosing to employ those units. It is still possible to modify ordinary stretchers in the field for any or all of the special purposes.

It is difficult to generalize about the utility of specialty units. Projects that make frequent use of special features (e.g, irregular cor-

TABLE 5-6 Specialty Unit Alternatives

Specialty units	Limitations	Advantages	Disadvantages
None	None	Standardization of units, low cost of units	Required field labor, potential variability of units
Some (sufficient to form 90° angles)	None		
Many	None	Field labor savings, consistency of units	Multiple types of units, cost of units

TABLE 5-7 Unit Assembly Alternatives

Unit assembly	Limitations	Advantages	Disadvantages
Pre-	None	Less user labor, consistency of units	More storage space (generally), some reduced flexibility
User	None	Less storage space, flexibility, cost	More user labor, potential variability of units

ners, brick ledges) will be likely to benefit more from having the specialty units designed to provide those features.

Unit Assembly

ICF units come either preassembled or with loose parts for the user to connect to one another. The unassembled parts are most often ties and face shells. Connecting them sometimes requires drilling holes or cutting notches in the foam. Insulform and Reddi-Form are preassembled except for their optional fastening strips, which are inserted by the user. Amhome panels are cut from standard foam sheet by the user.

The advantages of preassembly (summarized in Table 5-7) parallel those of specialty units. Less labor by the user is necessary, reducing the time and cost and the possibility of variation in the units from this source. But field-assembled units require less storage space because they are shipped "collapsed" into their components. However, preassembled units that fold flat (Folding Lite-Form and Fold-Form) take up nearly as little space. User assembly may be more flexible because one can put the unit together slightly differently from specifications to suit unusual circumstances. The materials cost of user-assembled units may be lower.

Research Reports

Research reports (also termed *evaluation reports*) are documents of about 5 to 15 pages issued by the model code organizations to describe recommended installation and inspection of proprietary construction products or systems. Product manufacturers compile the results of tests to show that their products meet all requirements of one of the code organizations. Working from these documents at the request of the manufacturer, the code organization determines whether the testing is indeed complete, and if so, writes a report. It testifies to the adequacy of the testing and the product's compliance with the requirements of the organization's code, describes the product and sometimes the correct installation procedure, and usually lists the steps required for an adequate inspection of an installation using the product.

Research reports are directed primarily at building department officials who are unfamiliar with a particular product or system. They acquaint the official with the essentials and provide a checklist for inspection. But reports are also of interest to designers, builders, and buyers because they provide a concise technical profile of the product and independent assurance that it meets a broad range of technical and structural standards.

Three U.S. code organizations issue reports for structural systems: the Building Officials Conference of America (BOCA), the Southern Building Code Congress International (SBCCI), and the International Conference of Building Officials (ICBO). BOCA publishes the Basic Building Code, which is the model upon which most state and local governments in the northeastern United States base their building codes. SBCCI publishes the Standard Building Code (the code adopted predominantly among governments in the Southeast), and ICBO the Uniform Building Code (predominant in the West). Canadian reports are issued by the National Research Council of Canada through the Canadian Construction Materials Centre (CCMC) of the Institute for Research in Construction. They are commonly referred to as CCMC reports.

Table 5-8 summarizes the considerations in deciding whether to favor an ICF that has research reports. The existence of a research report is virtually never an absolute requirement for an ICF building project. However, it provides building professionals the information and assurances outlined. The reports also frequently ease the approval process with local building departments. Without a report, officials are more likely to ask for more additional documentation: manufacturers' manuals, test reports, and the like. They are also more likely to re-

TABLE 5-8 Research Report Alternatives

Research reports	Limitations	Advantages	Disadvantages
Local	None	Acceptance by building department, assurance of requirements met, flexibility on stamping of plans	
Nonlocal	None		
None	None		Potentially more time with building department, stamping of plans more frequently required

quire that the plans be stamped by a professional, or that the stamp be that of a registered engineer (not an architect). Getting the plans for a house engineered and stamped can cost several hundred dollars. However, many manufacturers who have not yet received a report for their products have assembled substantial testing and convenient summaries of the results that may serve the same purpose. At the opposite extreme, a local official may choose to reject the use of a system even though it does have appropriate research reports.

If a particular ICF holds a research report from a code body not recognized locally (such as when a system with only an ICBO report is used in a southeastern town that bases its code on the SBCCI code), the reaction of local officials may vary. They may value the report as highly as one from their own model code organization, particularly since the model codes' requirements are generally similar. Or they may discount it.

In practice some building inspectors and professionals feel it less important to have research reports on flat systems. They recognize that a wall made with a flat system is virtually identical structurally to a conventionally formed concrete wall. Therefore the established rules for cast-in-place concrete can apply, and they may require no additional documentation.

It is difficult to predict the reaction of a particular building department to the presence or absence of a research report. Therefore it is safest to check with your local officials before selecting a final system.

Chapter

6

Building Envelope

Designing an ICF envelope involves about 10 major decisions. Separate sections in this chapter cover each of them. Where useful, we include a table of the scope, limitations, and advantages and disadvantages of each option to the decision, as in the previous chapter.

Important Note on Format

The notational and detailing conventions used throughout this chapter are introduced in one of its earliest sections, "ICF Levels." Therefore, regardless of one's interests, that section provides important background.

One noteworthy convention concerns the classes of systems used in wall section details. The "ICF Levels" section begins with three separate details, one depicting construction with a flat ICF system, one with a grid system, and one with a post-and-beam system. Details showing grid systems generally provide adequate information for design with any of the three. Therefore later details show grid walls only. Grid and post-and-beam design are slightly more constrained than flat because any feature embedded in the concrete must generally be positioned in the center of a cavity that is to be filled with concrete. Drawings using grid systems show this condition. Readers using flat systems may adapt the drawings by understanding that this constraint does not apply to them. Those using post-and-beam systems must recognize that this condition does apply, and in particular the concrete embedments must be in a cavity that is specified to be *filled*.

Dimensioning

The designer may size the structure so that key dimensions are multiples of the ICF's unit dimensions. This is a common practice in concrete masonry. Establishing wall lengths that are multiples of 16 inches (the

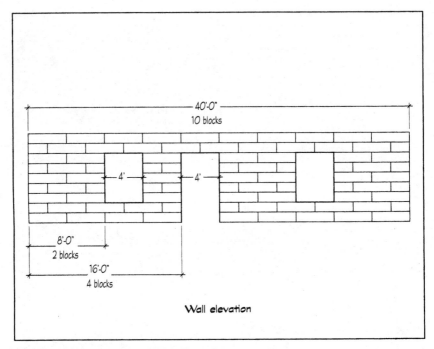

Figure 6-1 Dimensioning an ICF wall to minimize cutting.

length of a standard concrete masonry unit) and heights that are multiples of 8 inches (standard CMU height) minimizes the amount of cutting of blocks on site. Note that the relevant wall dimensions include not only total wall lengths and heights, but the dimensions of any section of wall that is interrupted as well. Thus the size and placement of openings is, ideally, also determined by the unit dimensions.

Figure 6-1 demonstrates how dimensioning could be accomplished with units from a block system of medium size. However, because the vertical seams of ICF units are generally staggered across courses, vertical breaks must cut through blocks on alternating courses at least. To facilitate dimensioning, some ICF manufacturers include in their manuals tables of wall heights and lengths that are consistent with the size of their units.

Table 6-1 summarizes considerations in the decision of whether to dimension plans. In practice the most common approach is not to do it at all. In this way no special planning of dimensions is necessary to accommodate the size of the particular ICF unit and no new constraints are placed on the elevations. Moreover, this leaves the field crews more flexibility to adjust dimensions should other design changes or unforeseen circumstances favor that. Conversely, the crews will have to do more cutting of formwork and will be likely to

TABLE 6-1 Dimensioning Alternatives

Dimensioning	Limitations	Advantages	Disadvantages
None	None	Less preplanning, assembly flexibility	ICF cutting, more scrap, low dimensional discipline
Heights	Particular wall and opening heights	Little horizontal cutting	Some preplanning, some loss of flexibility
Heights and lengths	Particular wall and opening heights, particular wall lengths and opening positions	Less cutting, less scrap, high dimensional discipline	More preplanning, less dimensional flexibility

generate more scrap. The discipline of plans that require breaks at modular locations will also be lost.

The opposite extreme—dimensioning both vertically and horizontally—is limited to designs with dimensions in standard increments. Its advantages and disadvantages are a mirror reflection of those associated with no dimensioning. Frequently the advantages of dimensioning are lost in practice when changes or eventualities dictate adjusting dimensions in the field.

Many designers settle for dimensioning certain heights only. Most systems are sized so that their units stack to conventional wall heights (e.g., 8 feet), and the levels of walls or of window sills rarely need to be adjusted in the field.

In general, the value of dimensioning is greater the more houses will be constructed from the same plans without variation. The time and cost savings for each house will eventually accumulate to outweigh the extra initial effort. Moreover, any necessary adjustments in the plans should appear while crews construct the first few houses, allowing the designer to correct and realize the full benefit of dimensioning in the later houses.

Wall Thickness

Many systems are available in alternative unit widths. This gives the designer some choice over how thick the final wall will be. The width is also associated with concrete content: for most systems, wider units have wider cavities, so the final concrete cross section will be greater as well.

Table 6-2 contains unit widths and concrete thicknesses for the systems. The concrete thickness reported is taken at the concrete's thickest point, since the cross sections vary in grid and post-and-beam systems.

Design

TABLE 6-2 ICF Width Data

System	Unit width (inches)	Concrete thickness* (inches)
Flat panel systems:		
Lite-Form (preassembled)	12	8
R-FORMS	8	4
	10	6
	12	8
	14	10
Grid panel systems:		
ENER-GRID	8	4.5
	10	6.2
	12	6.2
RASTRA	8½	4
	10	6
	12	6
	14	6
Post-and-beam panel systems:		
Amhome	9⅜	5½
ThermoFormed	8	5
Flat plank systems:		
Diamond Snap-Form	8	4
	10	6
	12	8
	14	10
Lite-Form (nonassembled)	12	8
Polycrete[†]	10⅝	5⅝
	11	6
	12⅝	7⅝
	13	8
	14⅝	9⅝
	15	10
	16⅝	11⅝
Quad-Lock	8⅛	3⅝
	10⅛	5⅝
	12⅛	7⅝
	14⅛	9⅝
Flat block systems:		
Blue Maxx	11½	6¼
	12⅝	8
Fold-Form	8	4
	10	6
	12	8

*Thickness of the concrete at its thickest point.
[†]Listed are standard widths. Other widths are available on a custom basis.

TABLE 6-2 ICF Width Data (*Continued*)

System	Unit width (inches)	Concrete thickness* (inches)
GREENBLOCK	9.88 (250 mm)	5¾
SmartBlock VWF	8	3¾
	10	5¾
	12	7¾
	14	9¾
Grid block systems:		
I.C.E. Block	9¼	6⅜
	11	8
Insulform	9.6	5¾
Modu-Lock	10	6½
Polysteel	9¼	6⅜
	11	8
Reddi-Form	9⅝	6
REWARD	9¼	6⅜
	11	8
SmartBlock SF 10	10	6½
Therm-O-Wall	9¼	6⅜
	11	8
Post-and-beam block systems:		
ENERGY LOCK	8	5
Featherlite	8	5
KEEVA	8	5

*Thickness of the concrete at its thickest point.
†Listed are standard widths. Other widths are available on a custom basis.

These dimensions dictate a few design details, as depicted in Figure 6-2. When ICF walls are built on a footing, it is usually desirable to center the cavities (and hence the concrete) over the footing. When a conventional concrete basement or stem wall is used, centering is also advantageous, but more critical is that the cavities not be so far off center as to extend any of the cavities beyond the edge of the wall below. Extending part of the foam face shell over the outside or inside edge of a concrete wall below is acceptable.

Table 6-3 contains considerations relevant to selecting a width. With the exception of the grid panel systems, the narrower the unit, the lower the concrete costs. The concrete cost will be roughly proportional to the concrete thickness as listed in Table 6-2. The units themselves may also be less expensive in narrower versions; however, this is not always so. Lumber used to form bucks around the perimeter of win-

114   Design

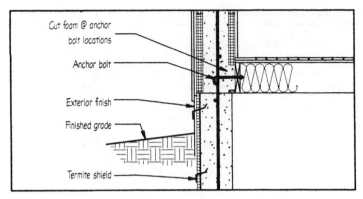

ICF centered on standard foundation with insulation

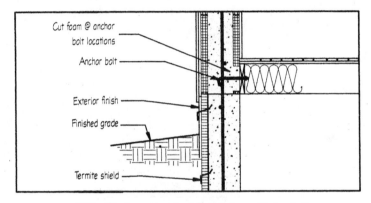

ICF offset on standard foundation with insulation

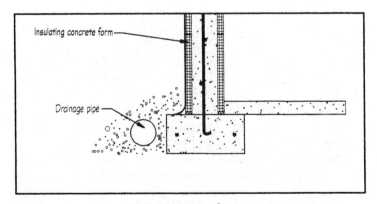

ICF centered on footing

Figure 6-2   Alternatives for aligning an ICF wall on a foundation wall or footing.

TABLE 6-3 Unit Width (and Wall Thickness) Alternatives

Unit width	Limitations	Advantages	Disadvantages
Narrow (8" or less)	Moderate wall strength	Lower concrete cost, narrower bridge lumber, more floor space available	Limited reinforcing clearance, more difficult span logistics
Medium (8 to 10")	None		
Wide (over 10")	None	Good reinforcing clearance, long spans easier, wider window sills	More concrete cost, wider bridge lumber, more floor space consumed

dow and door openings (see "Window and Door Mounting" later in this chapter) can also be of narrower dimensions, saving some money. The thinner wall also uses less footprint area. Thus in a constrained situation, a little more area is available for interior floor space.

Conversely, wider formwork makes setting rebar and placing concrete easier. It also leaves more room in the cavities for spacing rebar without resorting to vertical displacement. The space also allows concrete to flow more readily. The thicker cross section of concrete and the space allowing more reinforcement also enable the wall to be stronger where necessary, without extraordinary measures. Lastly, the depth of window sills can be greater. While preferences vary, frequently a deep window box is considered attractive and useful.

Wall Alignment

When adapting existing plans to ICF construction, one must decide how to align the exterior walls. ICF walls will almost always be thicker than those on the original drawings, so one cannot line up on both the exterior and the interior with the walls planned initially. The options, reflected in Figures 6-3 through 6-5 and in Table 6-4, are to align the interior surfaces of the new walls with the old, align the exteriors, or align the centers.

The most common practice is to align the interiors and allow the new exterior dimensions to be slightly greater. This will not be possible if it results in extending over legal setbacks or a physical barrier. Its major advantage is that no interior dimensions need be revised. This saves redesign time and is particularly significant when precise interior spaces have been allocated to accommodate specific features. In addition, there will be no loss of interior floor area. Occasionally the larger

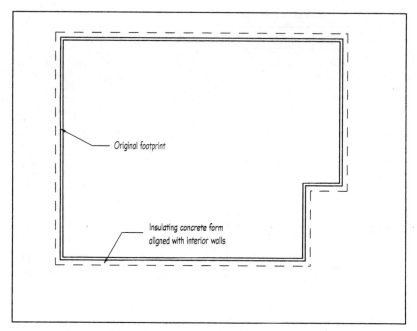

Figure 6-3 Aligning ICF walls with the interior of previously planned walls.

exterior could also require revision to exterior details, but these are usually not as critical as the interior ones. Roof and exterior finish areas will be slightly larger, increasing costs proportionately. Lastly, the perimeter of the footing or conventional stem wall or basement will need to be shifted out slightly to remain centered under the ICF walls.

The pros and cons of aligning the new and planned walls along their exteriors are the reverse of those of interior alignment.

Aligning the center of the new wall with the center of the originally planned wall has the one advantage of leaving the non-ICF parts of the foundation as they were. However, this is a small gain since redesign of the foundation is usually relatively easy. In most other respects this option represents the worst of both worlds. It is the least frequently used.

ICF Levels

It is possible to build only some of the stories of a house out of ICFs and the remainder of conventional materials. The logical alternatives are all levels of ICFs, the basement or stem wall foundation (as appropriate) of ICFs and above-grade walls of frame, a conventional

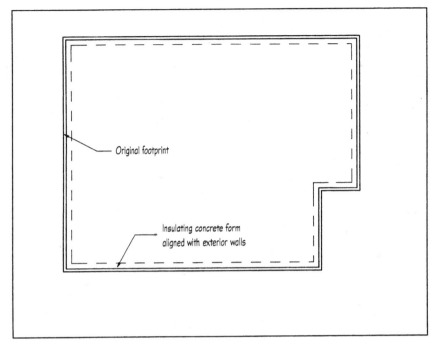

Figure 6-4 Aligning ICF walls with the exterior of previously planned walls.

concrete (masonry or cast) basement and ICFs above grade, or the first floor of ICFs with a conventional foundation and frame upper stories. Table 6-5 lists considerations applicable to each.

All-ICF

Building all levels with ICFs provides the benefits of these walls (noted in Chapter 1) along the entire house elevation. In addition, using the system from the footing up eliminates the time and costs of transitioning between materials.

Figures 6-6 through 6-8 depict a house with all ICF walls, including a basement. They feature a flat, grid, and post-and-beam system, respectively. Conventional groundwater provisions are recommended by most manufacturers: dampproofing or waterproofing on the exterior surface below grade, extending over the joint with the footing, and drainage around the footing perimeter. Above grade and extending a few inches below is a finish for protecting the foam and for aesthetics. (See "Exterior Finish" in Chapter 7.) The foundation wall connects to the footing with steel reinforcing bars. When the wall is grid or post and beam, these bars must align with the center of a filled vertical cavity.

118 Design

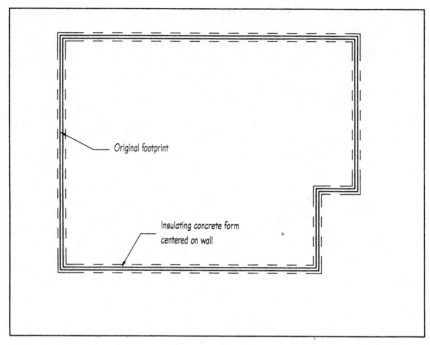

Figure 6-5 Centering ICF walls over previously planned walls.

TABLE 6-4 Wall Alignment Alternatives

Wall alignment	Limitations	Advantages	Disadvantages
Interior lines	Situations where exterior dimensions not restricted	Interior dimensions unchanged, interior space unchanged	Slightly higher cost, altered exterior details, altered foundation
Exterior lines	Situations where interior dimensions not restricted	Slightly lower cost, exterior dimensions unchanged	Altered interior dimensions, reduced interior space, altered foundation
Center lines	Situations where exterior and interior dimensions not restricted	Foundation unchanged	Altered interior and exterior dimensions, slightly reduced interior space

TABLE 6-5   ICF Level Alternatives

ICF levels	Limitations	Advantages	Disadvantages
All	None	Full benefit of ICF walls, efficiency of a single system	
All but roof ends	None	No need for angled pour	Inefficiency of multiple systems, potentially lower strength
Basement only	None	Ease of basement construction, ease of finishing basement	Lost ICF benefits above grade, materials cost of basement
Stem wall only	None (wherever stem walls are acceptable)	Ease of stem wall construction	Lost benefit of ICF walls above grade, materials cost of stem walls
All above grade	Heavy foundation	Benefit of ICF walls in living areas	Difficulty of finishing basement (if any), heat loss through basement
First floor only	Heavy foundation	Benefit of ICF walls on one floor	Inefficiency of multiple systems, lost benefit of ICF walls on other floors, heat loss through basement

Some code jurisdictions require an interior fire barrier over all foam, including in basements and garages, regardless of whether or not they are inhabited. (See Chapter 4 for a discussion of fire issues.) The ordinary method of satisfying this requirement is with gypsum wallboard.

Note that top plates and floor ledgers (see "Floor Decks" in Chapter 7) are in contact with the concrete. Local code may require, or the designer may prefer, to make them moisture-resistant. This can be accomplished by specifying pressure-treated lumber, or kiln-dried lumber wrapped at its contact points with a water-resistant barrier such as tar paper or plastic sheet.

For reference, Figure 6-9 shows the simplest variant of an all-ICF home: a single story on a slab foundation. The ICF face shells may align with or slightly overhang the exterior edge of the slab.

All ICF except roof ends

When building a gable or gambrel roof, some contractors construct frame roof ends on top of the ICF walls, as illustrated in Figure 6-10. The major advantage is that they can avoid a concrete pour into the end walls, which requires forming a slope along the top. This is pri-

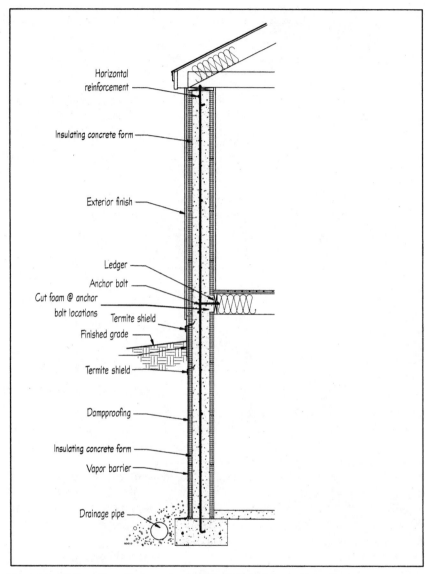

Figure 6-6 Wall section of a house built with all flat ICF exterior walls.

Building Envelope 121

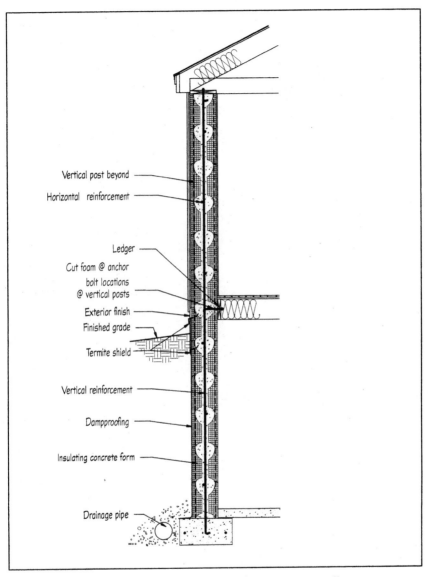

Figure 6-7 Wall section of a house built with all grid ICF exterior walls.

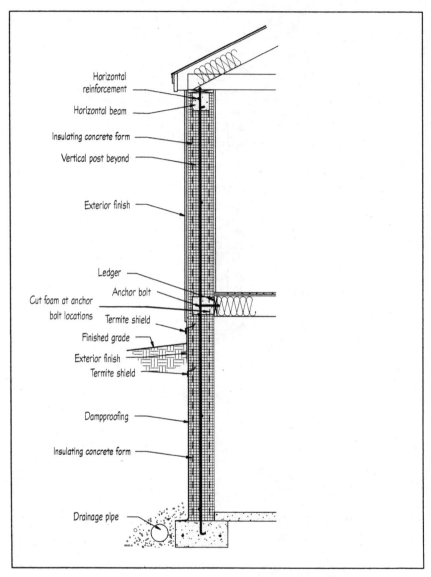

Figure 6-8 Wall section of a house built with all post-and-beam ICF exterior walls.

Building Envelope 123

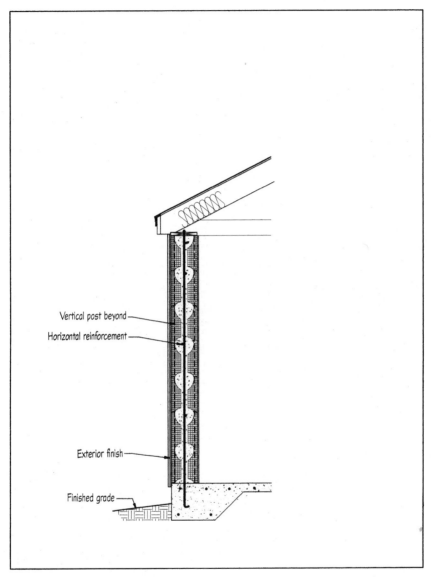

Figure 6-9 Wall section of a single-story house on grade built with grid ICF exterior walls.

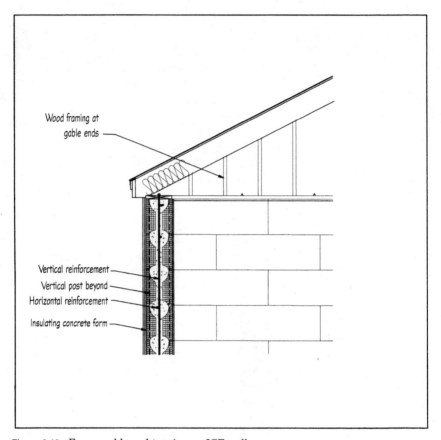

Figure 6-10 Frame gable end topping an ICF wall.

marily a concern with inexperienced crews. While angled pours involve some extra planning the first time, they are not particularly difficult. (See Chapter 15 for logistics.)

There are also a few disadvantages to frame roof end walls. The extra transition to a new material is inefficient. The end walls will not generally be as strong as ICF versions. This is most important in very high-wind areas, where gable and gambrel ends are frequent casualties in severe storms (as discussed in Chapter 4). Lastly, if the interior between the end walls is to be living space, the other benefits of ICF walls will be lost there.

ICF basement only

In some building projects ICFs are viewed simply as an alternative to a conventional basement wall system. Their advantages relative to conventional masonry or cast concrete basements are ease of construction, superior insulation, and readiness for complete finishing. Installing ICFs is generally easier and faster than other forms of basement construction. Frequently the carpentry crew that will build the above-grade structure can do it, eliminating one crew transition. Once completed, all that is usually needed to finish the basement is to run utility lines and install and finish wallboard. A conventional basement would require the additional steps of furring and insulating. Materials costs of ICF basements may be higher.

If ICFs are used only for the basement, one loses their benefits in the primary living space. In addition, crews must make a transition between wall systems.

Note in Figure 6-11 that most of the details that apply to a basement in an all-ICF house apply in this case as well. At the top of the wall comes a wooden top plate, as with a conventional concrete basement. Most often this is held in place with anchor bolts or metal straps.

Stem wall only

Upon first consideration, building a stem wall foundation of ICFs and all above-grade walls of frame appears illogical. As there is no living space between stem walls, no one will benefit from many of the advantages of ICF walls. Moreover, the materials cost is higher than that of conventional, uninsulated concrete wall systems.

Still, the ICF stem wall can be advantageous for logistical reasons. This is particularly true when building with a small workforce or in remote areas. The carpentry crew—or, if necessary, a single person—can build such a short ICF wall without formwork or mortar. The savings in time, labor, and coordination, compared with bringing in a separate crew to construct 3 to 4 feet of wall, can be decisive. In addition, insulating the stem wall can be more efficient than insulating the floor. Some localities now require it. The ICF accomplishes this without an extra step.

Note in this case as well that stopping ICF construction with the foundation fails to capture the benefits above grade. The stem wall detail drawn in Figure 6-12 is analogous to that of a basement.

All above-grade walls

Some projects forego the ICF basement/stem wall. This saves some money if the basement is not to be finished. Should the basement ultimately be finished, however, this approach leaves extra steps to be performed. It also misses an opportunity to eliminate the usual transitioning details and logistics between foundation and first floor. And

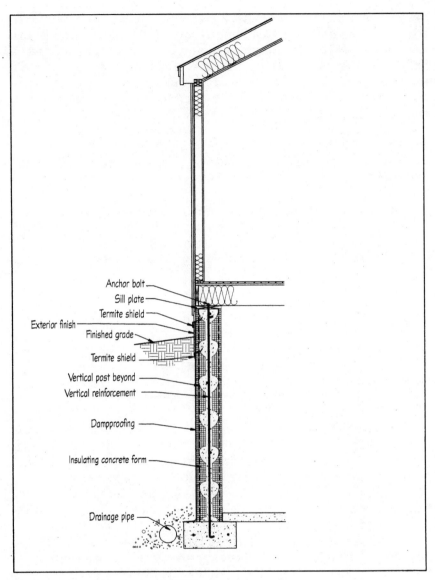

Figure 6-11 Frame house atop an ICF basement.

Building Envelope 127

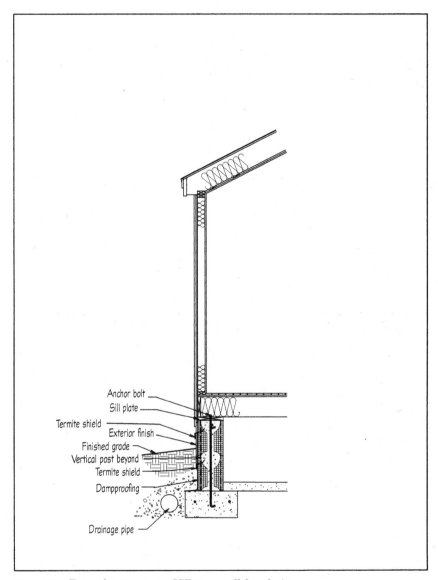

Figure 6-12 Frame house atop an ICF stem wall foundation.

conventional basements are rarely as well insulated. This is important because heat loss through basement walls can be substantial.

Bear in mind that the foundation must be strong enough to support a concrete structure above. Conventional concrete basements are ordinarily sufficient, but there is cause to check some foundation systems. Figures 6-13 and 6-14 contain details.

ICF first story only

There are few reasons to build only the first floor of a multistory home of ICFs (Figure 6-15). Some homes have a masonry structure on the first floor and frame above. However, this is primarily because the weight of the masonry and mortar causes higher costs and logistical difficulties in upper stories. These problems do not apply with lightweight ICF formwork. Among the negatives, a single story of ICFs in a multistory house maximizes the inefficiencies of transitioning between systems, denies the ICF benefits to the upper stories, and usually results in a less well-insulated basement.

One advantage to framing a second story will apply in rare cases. As discussed in Chapter 1, spans and projections are often constructed of frame. If the upper story is to have a large number of them, it might be easier to build it entirely of frame rather than a patchwork of two systems.

Insect Provisions

Although there are no confirmed reports of insects infesting ICF houses, some preventive measures are available for those who wish to be cautious. (See Chapter 4 for background.)

Subterranean termites

When the ICF extends into the ground, one or two breaks in the foam are a cautious practice in areas prone to subterranean termites (see Chapter 4). These force the insects into the soil, which can be treated with insecticide, or into the air, where pest control operators (PCOs) can detect them.

The common alternatives are a *termite shield* and a *foam gap*. Figures 6-16 and 6-17 show alternative termite shield arrangements; Figure 6-18 shows the foam gap. A termite shield consists of galvanized steel sheet (such as flashing) around the perimeter, cast into the concrete along one edge and extending through a seam or cut line in the exterior face shell. Its outer edge should extend 4 to 6 inches beyond the exterior ICF surface. This protruding edge may be bent over and glued flat to the surface and finished over, or allowed to extend beyond

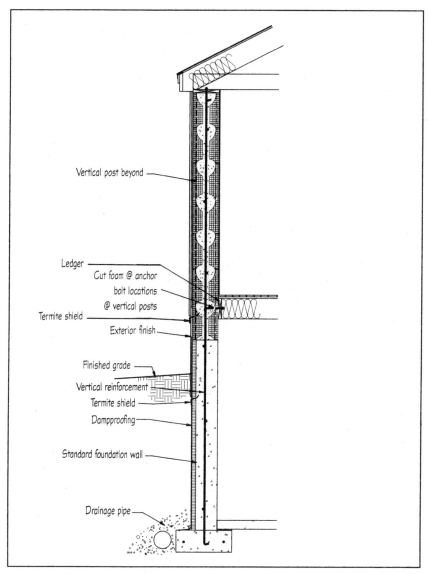

Figure 6-13 ICF house atop a conventional concrete basement with a floor deck supported by a ledger.

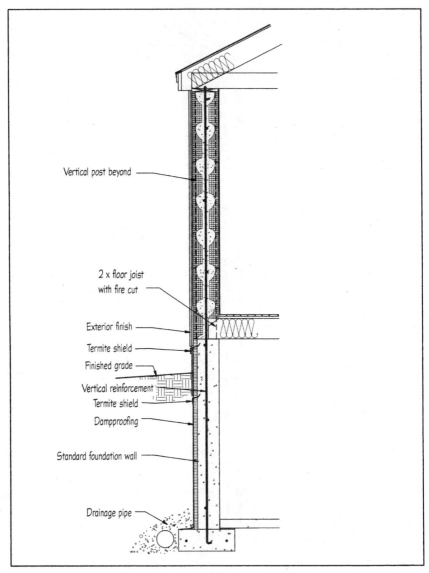

Figure 6-14 ICF house atop a conventional concrete basement with a pocketed floor deck.

Building Envelope 131

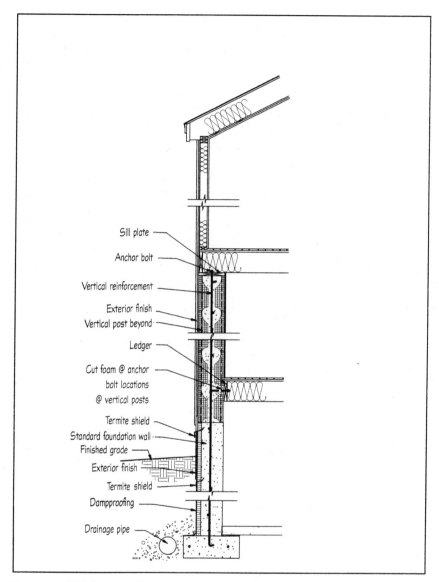

Figure 6-15 ICF first story between a conventional concrete basement and a frame second story.

132   Design

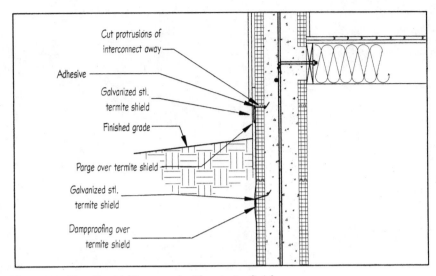

Figure 6-16 Termite shields covered with exterior finishes.

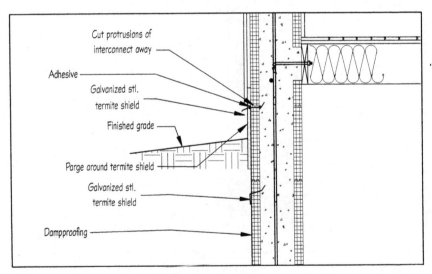

Figure 6-17 Termite shields protruding through exterior finishes.

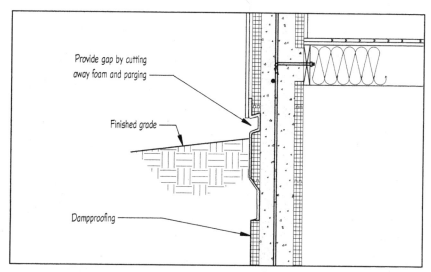

Figure 6-18 Foam gaps in an ICF foundation.

TABLE 6-6 Alternative Subterranean Termite Provisions

Provision	Limitations	Advantages	Disadvantages
Termite shield	At beam level	Retention of insulation, straightforward finish	Extra materials, construction logistics, does not meet some codes
Foam gap	At beam level	Low materials requirements, simple construction, required by some codes	Loss of insulation, creation of shelf, difficulty of finishing

the finish. The latter might be preferred if the finish material is not expected to be durable over metal. The foam gap is simply a 2- to 6-inch horizontal cutaway band from the exterior face shell that extends around the foundation perimeter. Note that with grid or post-and-beam systems, either form of break must be placed along a concrete beam.

Table 6-6 lists alternative considerations. The shield preserves most of the foundation insulation, whereas a gap eliminates a substantial section. Shields also may provide less opportunity for penetration by the elements; gaps add a horizontal ledge to the exterior face of the foam that is vulnerable to damage and water and dirt accumulation, and may be difficult to finish effectively. Conversely, the gap requires no extra material. It is also easy to control for quality. An inspector cannot be certain that a termite shield actually penetrates all of the foam, while a gap makes the concrete visible. The North Carolina State building code now requires two foam gaps on new construction, one 4 inches wide below grade to force termites into the soil (sometimes called the *treatment gap*) and another 2 inches wide above grade to facilitate inspection (sometimes called the *inspection gap*).

Carpenter ants

Recommended treatments for carpenter ants are the same as they are on wood frame. If one wishes to incorporate preventive measures, PCOs commonly recommend dusting the above-grade walls with a powdered insecticide just before finishing them. In the case of an ICF, this translates to dusting all exposed foam surfaces, inside and out.

General steps

When the need for preventive measures is considered high, one may wish to place a priority on constructing all levels of concrete (see "ICF Levels" in this chapter) and using only pressure-treated lumber in bucks (see "Window and Door Mounting" in this chapter), top plates (see Chapter 7), and floor deck ledgers (see Chapter 7).

Spans and Projections

As discussed in Chapter 1, upper-level walls that are unsupported from below raise special considerations. Spans over interruptions in a running wall, such as a window or door opening, are routine with ICFs. We discuss in Chapter 10 the sizing of lintels to accommodate them. Somewhat different are entire walls that span open space (such as a second-story wall that is set back), entire walls that project beyond the perimeter of the story below (such as a second-story garrison), or narrower projections designed to cantilever beyond the structure below (such as bays and dormers). In these instances the weight of the ICF walls above requires alternative support.

Building the upper walls out of ICFs is a valuable option done with increasing frequency. As listed in Table 6-7, this approach retains the benefits of ICFs in the affected second-story walls and avoids transitioning between materials. However, as this is not yet routine, it requires careful engineering for each project. Alternative structural methods for supporting upper-story ICF walls are discussed in Chapter 9. It may also require temporary bracing below to hold up the formwork during the pour.

Details for frame spans and projections are presented in Figures 6-19

TABLE 6-7 Span and Projection Alternatives

Span or projection	Limitations	Advantages	Disadvantages
ICF	Uncertain	Retain ICF benefits, potential strength and clearspan distance	Novel engineering, temporary bracing (possibly)
Frame	Same as conventional	Similarity to current practice	Inefficiency of multiple systems, loss of ICF benefits
Elimination	Constant perimeter	Ease of construction, elimination of structural dilemma	Loss of design flexibility

through 6-21. The frame approach has the advantage of familiarity, with the loss of ICF benefits in the affected walls and some inefficiency from switching between wall systems. Setback span walls require a beam or posts below for support if they are long. Projections rest on cantilevered floor joists, and all walls of the projection (side and front) are frame. Where a sidewall is a linear continuation of an ICF wall (e.g., the sidewalls of a garrison that runs the full width of the house), it is necessary to frame it to match the wall thickness of the ICF. This is most often accomplished with top and bottom plates matching the width of the ICF units and 2×4 studs offset to be alternately flush with the inside and the outside.

Modifying plans to eliminate spans and projections eliminates the need for any special structural provisions or materials decisions. This approach obviously limits one to houses with walls that have consistent perimeter dimensions up the entire elevation.

Division of Openings

When multiple door or window openings fall close together, the designer has the option of separating them with a narrow section of ICF wall or with frame. Table 6-8 compares the considerations, and Figures 6-22 and 6-23 contain details.

The ICF between openings forms a structural column of great strength. It also saves some transitioning between systems. Note, however, that in most applications the crews will build a wooden buck for each opening anyway (see "Window and Door Mounting"). A section of frame wall can be readily incorporated into one large buck that encompasses both openings. And when the wall section is narrow, the ICF

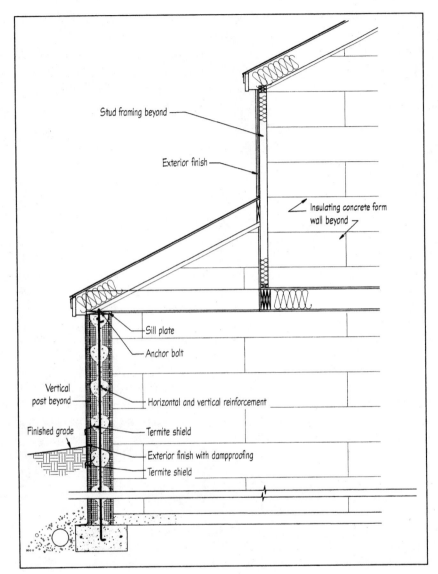

Figure 6-19 Setback upper-story frame wall atop lower-story ICF walls.

Building Envelope 137

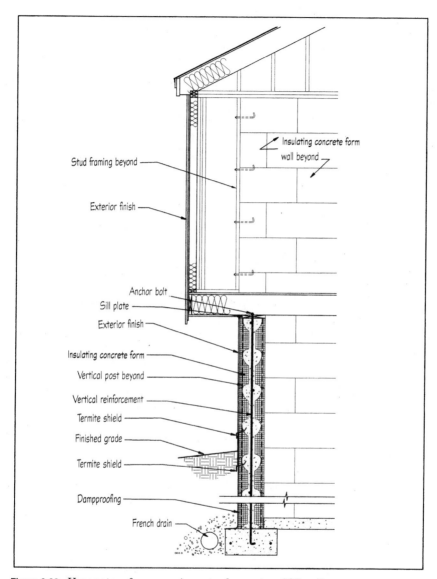

Figure 6-20 Upper-story frame garrison atop lower-story ICF walls.

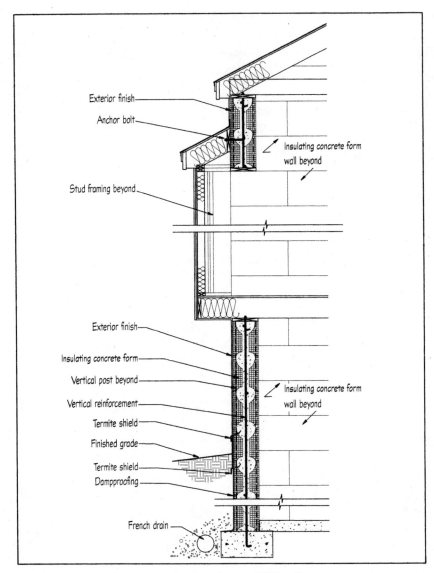

Figure 6-21 Narrow frame projection from house of ICF walls.

TABLE 6-8 Opening Division Alternatives

Opening division	Limitations	Advantages	Disadvantages
ICF	One cavity width	Efficiency of consistent system, strength	(Sometimes) cutting and waste
Frame buck	Certain span lengths	(Sometimes) less cutting and waste	Inefficiency of multiple systems, greater span for lintel

also involves more cutting and waste than usual. In any case, an ICF should not be used when the space involved would not include at least one complete vertical cavity. In a narrower wall section there will be insufficient space for reinforcement, and placing concrete will be difficult.

Using frame, conversely, saves some cutting and waste. But lumber is normally not strong enough to support the ICF wall above. Thus the reinforcement running in the ICF over the combined two openings will have to be strong enough to form the equivalent of one continuous lintel over one wide opening.

Window and Door Mounting

Table 6-9 lists alternative methods of mounting windows and doors into ICF walls, and Figures 6-24 through 6-26 provide details.

The *recessed buck* is so named because it is inserted into the cavities of the formwork. It is sometimes also referred to as a *stucco buck* because it is most commonly used when the exterior finish will be some form of stucco. It consists of a lumber frame sized to be a rough opening for the window or door, and of a width to insert into the formwork, flush with the edges of the ICF units, before the concrete pour. The recessed buck is held temporarily in place with adhesives or with insulation nails or screws through the foam. Protruding fasteners placed around its perimeter lock it into place in the concrete. Duplex nails hammered into the buck are normally sufficient fasteners; the heads embed in the concrete. For high strengths one can specify anchor bolts instead.

The recessed buck preserves the insulation of the ICF right up to the opening. It also leaves a continuous exterior foam surface, which is ideal for a later stucco finish. The lumber provides an adequate fastening surface for windows or doors that are attached with nails or screws through the frame, or the narrow-flange windows frequently used in masonry construction. (See "Window Details" later in this chapter.) Conversely, nailing or screwing ordinary flanged windows or other

140   Design

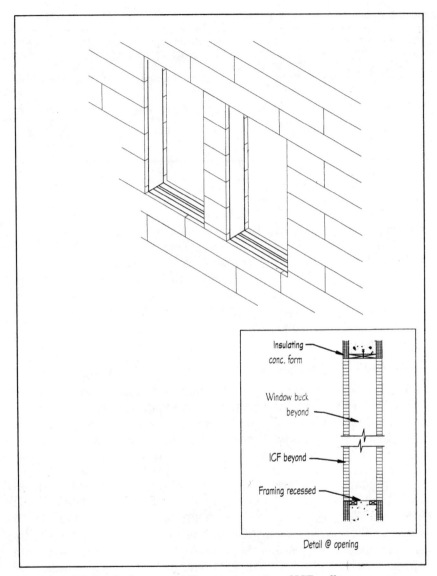

Figure 6-22 Window bucks separated by narrow section of ICF wall.

Building Envelope 141

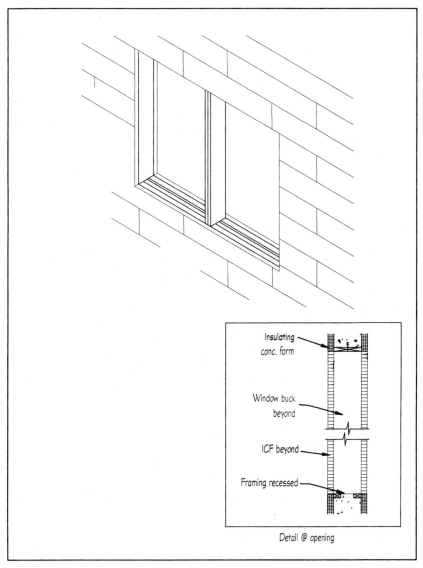

Figure 6-23 Single combined buck to accommodate two closely spaced window openings.

Design

TABLE 6-9 Window Mounting Alternatives

Mounting	Limitations	Advantages	Disadvantages
Recessed buck	None	Ease of stucco finishing, consistent insulation	Difficulty of attaching flanged windows, nailed and screwed sidings, solid trim
Protruding buck	None	Ease of mounting flanged windows, nailed and screwed sidings, solid trim	Slight loss of insulation, greater materials requirements
Channel buck	Availability of parts	Low cost, time saved	Limited experience

solid attachments (such as solid trim or sidings that are attached with fasteners) around the opening is more difficult with this buck.

The *protruding buck* differs in that the lumber's width matches that of the ICF unit, and it is placed entirely within the opening rather than being recessed into the cavities of the formwork. It is sometimes called a *flanged buck* for two reasons: it is outfitted with flanges front and back all around its perimeter before the concrete pour to reinforce the formwork around the opening, and it is most commonly used with wide-flanged windows and doors. Because it leaves lumber flush with the exterior, the protruding buck allows easy attachment of window flanges and trim nailed or screwed to the outside. It has the disadvantage of replacing a band of foam insulation around the window with wood (just as frame construction displaces insulation around openings and at studs and plates). It also requires extra material: flanges and wider buck lumber. But the flanges are temporary and reusable, except in special cases. (See "Exterior Finish" in Chapter 7.)

Some rules apply to both flanged and recessed bucks. The sill is normally constructed of two parallel pieces of lumber with a 3- to 4-inch gap in between. The gap is important for placing concrete in the portion of the wall under the sill. Local codes may require, or the designer may prefer, making the lumber moisture-resistant. This is most often accomplished by specifying that it be pressure-treated, or that it be wrapped with a water-resistant membrane (usually tar paper or plastic sheet) at the points where it contacts concrete. Although opinion is split on whether this is necessary for jambs and lintels, the majority of designers prefer pressure-treated for the sill material because of its potential exposure to slow-moving water. Note also that pressure-treated lumber is more insect-resistant. This is a consideration in areas of high pest risk.

Building Envelope 143

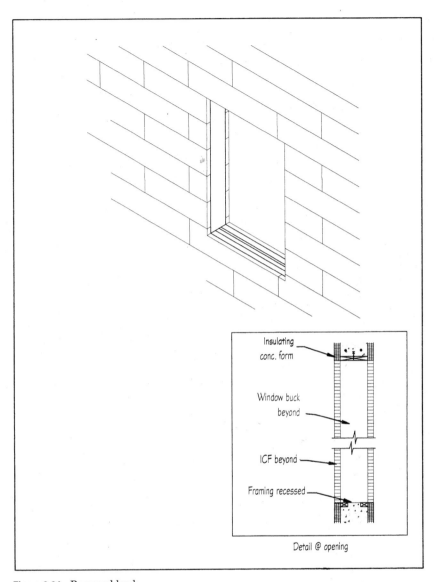

Figure 6-24 Recessed buck.

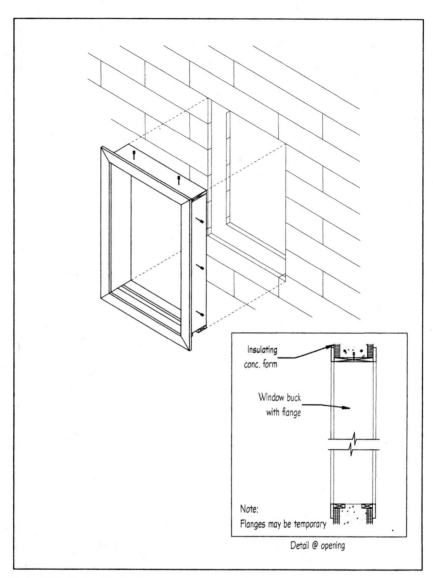

Figure 6-25 Protruding buck.

Building Envelope 145

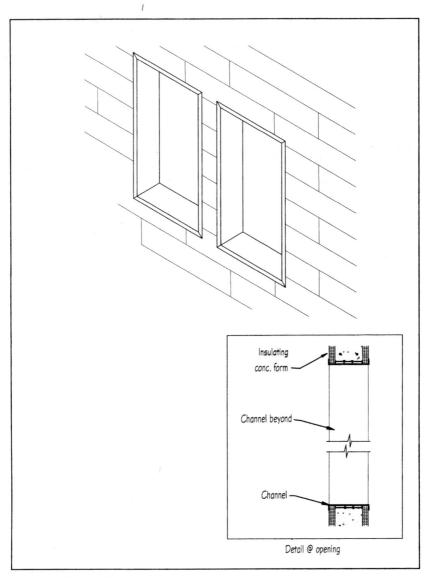

Figure 6-26 Channel buck.

Some ICF manufacturers are selling or developing a plastic or metal channel that replaces wood in the buck. Potential advantages are material and time savings. However, the channel buck is a recent development with limited field experience, and the materials are available from only a few manufacturers.

Window Details

Good window detailing on ICFs uses common materials in logical ways. However, the arrangement is sometimes different from conventional frame practice. Designing window mounting on houses to be finished with stucco in high-rain areas merits particular care because of the potential, with any wall system, for leakage at the joints between window and wall. (See Chapter 4 for background.)

Figures 6-27 through 6-31 contain details for common arrangements. When screw-through windows are used, the sealant provides the only barrier at the jamb and sill joints (Figures 6-27 and 6-28). The quality of the material and its application are therefore critical.

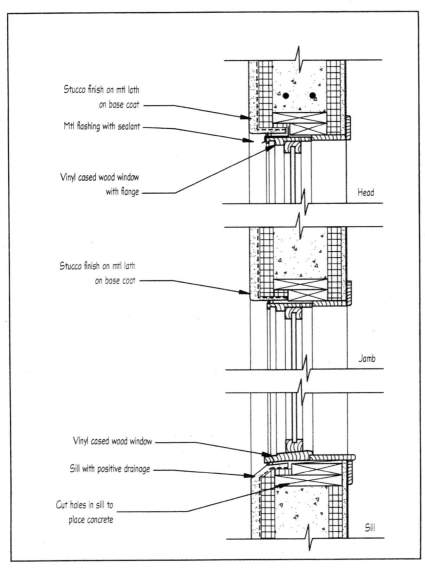

Figure 6-27 Screw-through window mounted in wall finished with stucco.

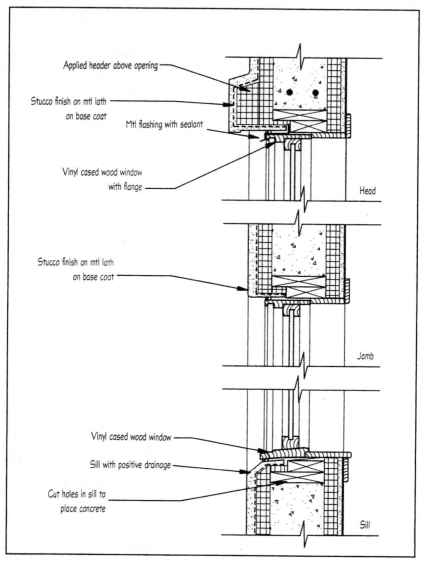

Figure 6-28 Screw-through window mounted in wall finished with stucco and foam header.

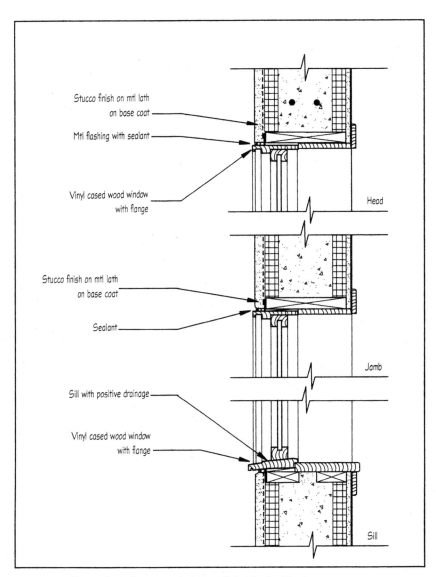

Figure 6-29 Flanged window mounted in wall finished with stucco.

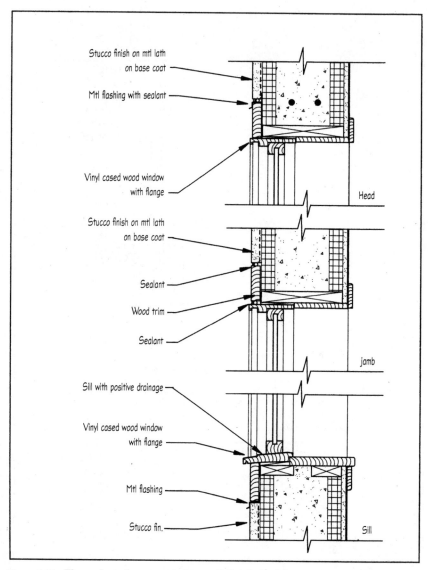

Figure 6-30 Flanged window mounted in wall finished with stucco and wood trim.

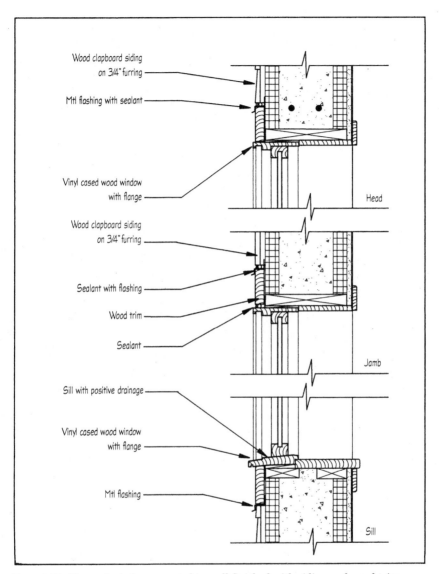

Figure 6-31 Flanged window mounted in wall finished with siding and wood trim.

Chapter 7

Attachments

Although ICF exterior walls leave the design of the rest of the house largely unaffected, connecting certain other components involves some unusual considerations. Most of the important connections can be handled in multiple ways, so for each of these we present the major alternatives and their limits, pluses, and minuses, as in previous chapters.

Floor Decks

Designing floor decks involves two decisions. The first is what material to use, the second is which of several alternative deck configurations to choose.

Material

Table 7-1 reviews the alternative materials. Most common for floors is frame. Framed floors are built and attached to ICFs much as they are

TABLE 7-1 Floor Deck Material Alternatives

Floor deck material	Limitations	Advantages	Disadvantages
Wood/steel frame	None	Cost, familiarity	Flex and squeaking, low mass, sound passage, susceptibility to fire
Reinforced concrete	None	Rigidity and strength, sound attenuation, span capabilities, fire resistance, thermal mass, aesthetic potential	Cost, unfamiliarity, thermal bridge

in frame houses, so they have the advantage of being familiar to the designers and crews. Their major disadvantage is their propensity to flex and squeak. They also have more limited spanning capabilities than concrete decks. Wooden decks may be deeper than concrete to begin with, and spanning increasing distances without support requires much faster increases in member size, and hence depth and cost.

The other option is a reinforced concrete floor. Concrete decks are relatively easy to use with ICFs because the concrete walls can bear the floor's greater weight without special provisions. In addition, the incremental logistics of using them are sometimes smaller because the crews are already working with concrete and reinforcing. Concrete floors are more rigid, stronger, and more fire-resistant than wood or steel frame. They span longer distances without increases in depth, and they can form a diaphragm that ties the ICF walls together into an unusually strong structure. Most varieties reduce the passage of sound between floors of the house sharply, with STCs about 10 points higher than for conventional frame decks. (See Chapter 1 for background on sound attenuation.) They also have greater thermal mass. This is particularly useful in applications that use interior storage or distribution of latent heat, such as passive solar and hydronic radiant heating. Some types of concrete floors can take advantage of the unique, distinctive finishes possible with exposed concrete. They also accept the same floor coverings used on frame decks.

Concrete floors usually cost somewhat more in the typical residential application. The differential declines or disappears as one moves to designs with very long clear spans, or as one attempts to outfit frame decks with the extra benefits that come as standard with concrete. Concrete floor decks will also be unfamiliar to many residential crews in the United States and Canada, and they generally create a greater penetration of the inside face shell of ICF formwork. (See the details that follow.) This might reduce the effective R-value of the exterior envelope slightly.

Frame decks

Table 7-2 covers alternative methods of building floor decks of frame, and Figures 7-1 through 7-3 provide details. These methods have been used primarily with joists of dimension lumber, and occasionally of steel. However, most of them are readily adaptable to engineered lumber, wooden I-joists, or wood or steel trusses.

Ledger. Perhaps most commonly used is the ledger. Under this method, rectangles or circles are cut out of the inside face shells of the ICF formwork at floor deck height. A ledger board holding anchor bolts

TABLE 7-2 Frame Floor Deck Alternatives

Floor deck	Limitations	Advantages	Disadvantages
Ledger	None	Long experience, good rigidity	Complex assembly, materials requirements
Pockets	None	Long experience, materials savings, labor savings	Inflexibility of joist placement
Embedded joist hangers	None	Labor savings	Inflexibility of joist placement, limited experience

is mounted over the holes. During the pour, concrete fills the rectangles, backing up the ledger and locking the bolts securely. The joists of the floor deck are attached to the ledger with joist hangers. On grid systems, if the ledger is placed at a level that falls between horizontal cavities, the cutouts are made at the vertical cavities. On post-and-beam systems it is best placed at the level of a beam. Careful practice dictates adjusting floor height or creating an extra beam as necessary to accomplish this. It may be necessary or preferable to specify the ledger boards as pressure-treated or to specify a water-resistant membrane between lumber and concrete. (See "ICF Levels" in Chapter 6.)

There is long, successful experience with the ledger. It produces a solid floor in a manner that is mostly familiar to carpenters. However, the many joist hangers add to materials cost and assembly time.

Pockets. Pockets are another popular alternative. Under this method the crew places the joists in position, ends in the formwork core, before the pour. (Chapter 16 covers logistics.) Metal anchors embed in the concrete and connect it to a joist every few feet. Required anchor spacing depends on the local code. The pocket method saves labor and material. (Ledgers, joist hangers, and nails are all omitted.) It holds the ends of the joists particularly solidly. The method also has a long track record in commercial construction.

Its major disadvantage is that the position of the joists is inflexible after the pour. Thus preplanning is necessary, and carpenters constructing the floor deck do not have their customary freedom to shift members to compensate for late design changes or unforeseen circumstances.

Some codes require that the joists of a pocketed floor be *fire cut*. This means that the top corners of their ends set in the pockets are cut at an angle to allow the joists to fall out of the wall (rather than

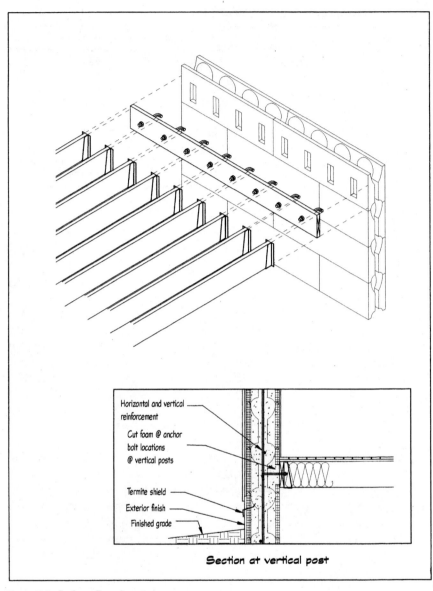

Figure 7-1 Ledger floor framing.

Attachments 157

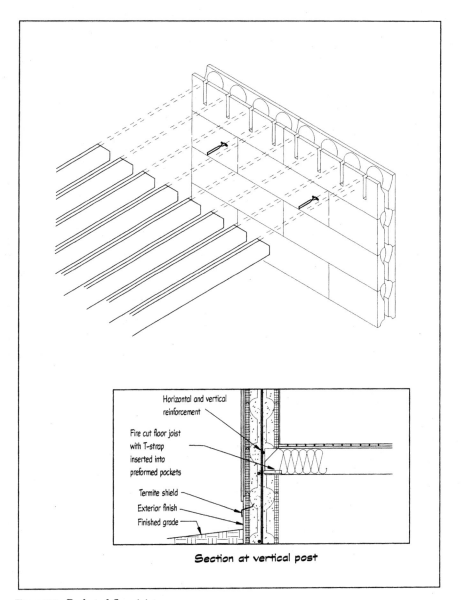

Figure 7-2 Pocketed floor joists.

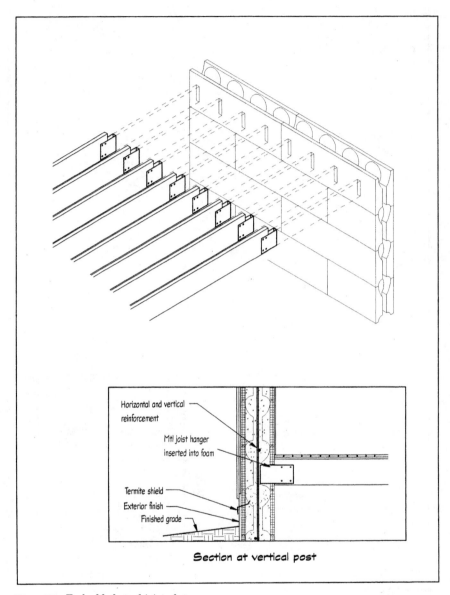

Figure 7-3 Embedded steel joist plates.

put severe forces on it) in the event that a fire burns through them or their support at the opposite end.

Pocketing results in contact between concrete and joist ends. Some codes require, and conservative designers may prefer, a water-resistant coating or membrane around the end of the joists to ensure against any potential moisture damage to the lumber. The same does not apply to nonlumber joists. Note also that some engineered wooden joist products already come with a coating.

Embedded joist hangers. Some ICF manufacturers sell special steel plates or joist hangers that one inserts into the formwork before the pour. They lock into the concrete. The crew then fastens the joists directly to them. They have the advantages of speed and ease of construction. However, like pockets they constrain joist location once concrete is cast. This is also a relatively new technique, with a limited record of experience.

Concrete floors

The concrete floor systems used thus far in ICF home construction fall into four categories: concrete on steel joists, concrete on steel deck, concrete slab and joist, and precast concrete. Figures 7-4 through 7-7 depict a representative version of each.

Concrete on steel joists. Light steel joists can support a steel deck or ribbed pan on which $2\frac{1}{2}$ to 4 inches of concrete is placed. Floors with a total depth of $10\frac{1}{2}$ inches (8-inch joists plus $2\frac{1}{2}$ inches of concrete) achieve spans of 20 feet without special measures. Deep decks (28 inches) readily span 40 feet.

A proprietary steel joist system named *Hambro* is well suited to residential construction and frequently used in ICF homes. Figure 7-4 contains a detail. The joists are special steel trusses. Instead of a permanent steel pan it uses sheets of plywood temporarily held up with roll bars that run crosswise between joists. Steel mesh is laid on top, then covered with concrete. The top chords of the joists become embedded and act as reinforcement. The plywood and roll bars are removed from underneath and can be reused. Because of the strength provided by the composite action between joists and concrete, this system can be somewhat lighter than designs of plain joists and steel decks.

Concrete on steel deck. Metal deck made from plain or galvanized steel sheet rolled into various ribbed profiles can be used to form concrete floor slabs. (See Figure 7-5 for an example.) For light construction, the material is generally made from 18- to 24-gauge sheets with $1\frac{1}{2}$- to 3-

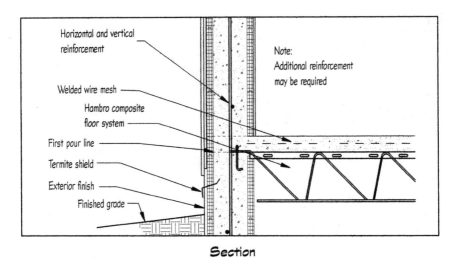

Figure 7-4 Hambro steel joist floor deck.

inch-deep ribs. Steel deck can serve strictly as a form for the slab, or it can be fabricated to bond with the concrete and act with it structurally.

When the decking serves solely as a form, the concrete slab must be sufficient to carry all loads. The concrete must therefore be specified and reinforced by added steel, as appropriate. Temporary supports installed under the deck at approximately 5 feet on center must remain in place until the concrete reaches 80 percent of its design strength.

For a composite steel deck floor, deck with "dimples" formed into the vertical flutes physically bonds to the concrete to perform the structural function of reinforcement. A 22-gauge deck 3 inches deep topped with 3 inches of concrete (resulting in a total floor depth of 6 inches) can span 15 feet with no additional reinforcement. With heavier deck or the addition of reinforcement, greater spans are possible. Manuals are available for the design of composite deck floors in both the United States (American Society of Civil Engineers, 1994; Heagler, Luttrell, and Easterling, 1991) and Canada (Canadian Sheet Steel Building Institute, 1988; Canadian Portland Cement Association, 1995).

Concrete slab and joist. Forms with deep ribs create floors that combine concrete slabs and joists in one continuous casting of material. Removable (temporary) metal or fiberglass forms are available, as well as steel or fiberglass forms that bond to the concrete to act as reinforcement. Typical systems create 8-inch-deep ribs, 2 to 3 feet on center. Additional concrete depth is 2 to 3 inches, for a total floor thickness of 10 to 12 inches. Usually one reinforcing bar goes near the bottom of each rib and wire mesh across the top surface. With composite (bonding) forms,

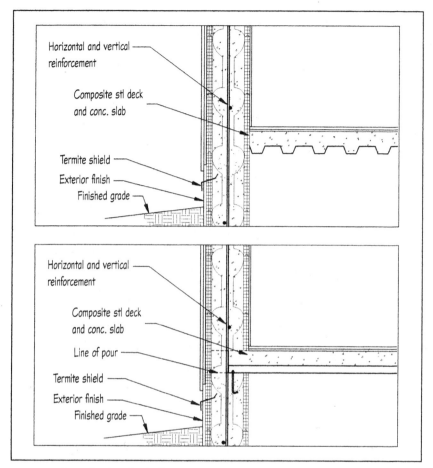

Figure 7-5 Front (top) and side (bottom) view of composite steel deck.

spans of 16 to 26 feet are achievable without special measures, depending on the depth of the floor and the reinforcement schedule.

Several manufacturers are preparing systems that use foam forms (instead of fiberglass or metal) to create concrete joist floors. Such systems have long been sold by European companies and installed in thousands of buildings abroad (Figure 7-6). Hollow-core foam planks with periodic ribs stay in place permanently. They arrive with small "joists" (usually of reinforced concrete) built in to hold them stiff during installation and pour. In addition, they are braced below with temporary supports every few feet. Like ICF wall systems, attachments for finishes can be incorporated into the forms. Plans call for many of these new systems to be designed specifically for use with ICF walls.

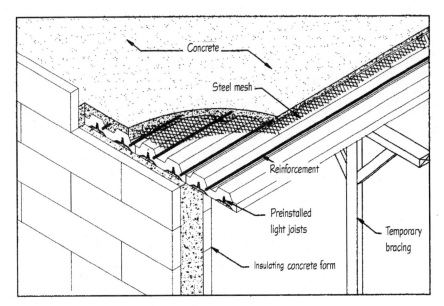

Figure 7-6 Cutaway diagram of a concrete slab and joist deck created with a foam form system.

Precast concrete. Precast prestressed hollow-core concrete slabs for housing (as detailed in Figure 7-7) are usually 4 feet wide and 6 or 8 inches thick. Most have evenly spaced voids ("cores") running the length of the slab with prestressing tendons in between. They are cut to final lengths at the plant. Widths and shapes may be customized for stair openings or other special situations. Holes for services are factory-cut or cut on site with concrete drills or saws. The slabs are set on ICF walls by crane and anchored with bent reinforcing bars set into the shear key joint between the planks. Grouting the joints completes the installation.

Precast products have constant depth with a relatively flat top and a smooth underside that requires minimum additional finish. A thin grout or concrete topping is required under thin flooring (such as vinyl).

New systems. Other concrete floor systems are likely to become available in the future. Candidates include several already proven in other parts of the world that are adaptable to North America.

Roofs

To date roofs on ICF houses in the United States have almost always been conventional wood or steel frame, although many in Europe employ concrete roof systems. There are three common methods for at-

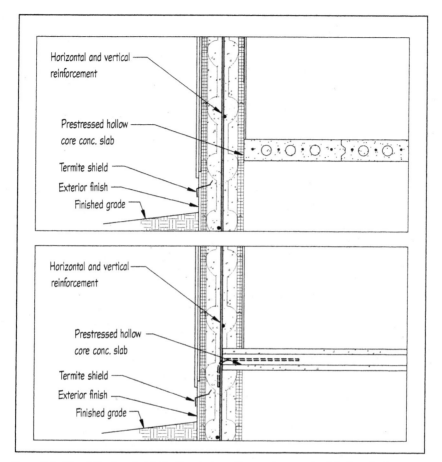

Figure 7-7 Front (top) and side (bottom) view of precast concrete slab floor deck.

taching frame roofs to the walls. These are compared in Table 7-3 and detailed in Figure 7-8.

Steel straps are most common in high-wind areas. Commonly called *hurricane straps,* the crew embeds one end of each into the concrete when filling the top story. There is one strap at the planned location of each roof member. When the roof structure is built, each strap is wrapped around and nailed to its respective member.

A strapped roof is extremely resistant to uplift, the most damaging force placed on a roof in wind storms. Also, its construction uses the least amount of material. However, this method leaves the crew little flexibility in the placement of their joist and rafters. At most they can safely move the member 1 or 2 inches to either side of its intended

164 Design

TABLE 7-3 Roof Connection Alternatives

Roof connection	Limitations	Advantages	Disadvantages
Straps	None	Resistance to uplift, low materials requirements	Inflexibility of roof member placement
Protruding top plate	None	Flexibility of roof member placement, ease of attaching trim	Need extra connectors to resist uplift
Recessed top plate	None	Flexibility of roof member placement, ease of stucco finishing	Need extra connectors to resist uplift, possible difficulty of attaching trim

strap. Advance planning is important, and late changes in roof design are more difficult.

The alternative is to anchor a wooden plate on top of the wall with anchor bolts. This leaves the crew the same flexibility they would have when building a roof atop conventional frame. Note that the plate need not be doubled (as often required in frame construction). The continuous support of concrete below ensures that it will not flex. Roofs on top plates can resist uplift as well as strapped roofs, but in most cases extra fasteners are required to accomplish this.

There are two types of plates, analogous in their properties to the protruding and recessed buck. The first is the *protruding top plate*, which sits above the top of the foam. The concrete is placed to fill the formwork exactly, and the plate goes on top. Ideally the plate is sized and set so that its edges are flush with both the exterior and the interior surfaces of the foam. This permits easy attachment of trim and wallboard. It could be made of narrower material and set flush with the outside, leaving the top inch or two of wallboard without backing.

The *recessed top plate* is usually more appropriate when the exterior is to receive stucco up to the roof. It is constructed of lumber sized to fit inside the cavities of the formwork, with its top face flush with the top edge of foam.

Interior Walls

The method of connecting the ends of abutting interior walls to the ICFs is normally left to the contractors. However, some designers may prefer to specify it to ensure a certain level of rigidity.

The common options are listed in Table 7-4 and detailed in Figure

Attachments 165

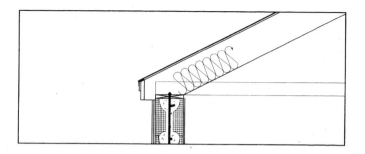

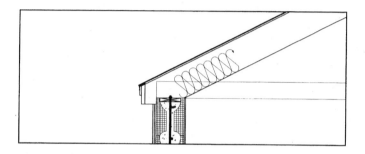

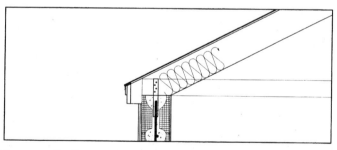

Figure 7-8 Roofing members attached with (top to bottom) protruding top plate, recessed top plate, and metal straps.

TABLE 7-4 Interior Wall Connection Alternatives

Interior wall connection	Limitations	Advantages	Disadvantages
Anchor bolt	None	High rigidity	Cost, inflexibility of placement
Straps	None	Good rigidity	Cost, inflexibility of placement
Duplex nails	None	Low cost	Temporary safety hazard, inflexibility of placement
Glue	None	Low cost, flexibility of placement	Uncertain rigidity
Concrete fasteners	None	Low cost, flexibility of placement	Long fastener required, uncertain rigidity
No connection	None	Low cost, ease and speed of assembly, flexibility of placement	Low rigidity

7-9. Generally speaking, the strongest connections result from embedding fasteners in the concrete. The strongest and most expensive of these options is the anchor bolt, followed by straps and then duplex nails oriented with the point out. All require the interior wall positions to be preplanned, offering little flexibility of repositioning after the pour. The points of the duplex nails are also somewhat hazardous until the wall studs are hammered over them. Temporarily covering them with small wooden blocks is advisable.

After the pour the end studs can also be glued to the foam surface. This has the advantage of unconstrained placement along the ICF wall, and in practice appears to form a rigid enough connection for most walls.

The stud ends can also be secured with concrete fasteners long enough to pass completely through the foam and into the concrete.

Some builders do not attach the ends of interior walls to the ICF walls at all, but merely nail them to the floor below and the framing above. This is flexible in placement and is the easiest and fastest method. However, some dislike the lack of a hard connection.

Utility Lines

There are several ways to run electrical and plumbing lines along ICF walls. Depending on the lines involved and designer preferences, the method used may be specified on plans or left to the discretion of the

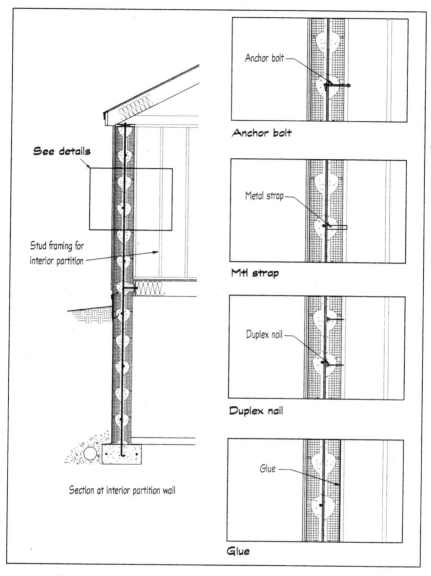

Figure 7-9 Interior walls attached to ICF walls with alternative fasteners.

crews. Table 7-5 lists the alternative methods, and Figures 7-10 and 7-11 show details.

One frequently has the flexibility to route lines through interior walls instead of exterior. This is preferred practice for large-diameter lines (such as drain pipes) regardless of the wall system used, as plac-

TABLE 7-5 Utility Line Placement Alternatives

Utility line placement	Limitations	Advantages	Disadvantages
Interior walls	Where interior walls are available	Familiarity	Potentially inconvenient routing
In foam	Small lines (up to 2" diameter)	Flexible routing of small lines	Unfamiliarity, slight loss of insulation
Between furring	Small lines, code-approved areas	Flexible routing of small lines, familiarity	Cost, shallowness of mounting
Fully in concrete	Where structurally permissible	Permanence of channel, security of lines, flexible routing	Cost, required advance planning, potential for weak points, materials incompatibilities, low accessibility for repair
Partly in concrete	Where structurally permissible	Permanence of channel, flexible routing	Cost, required advance planning, potential for weak points, some loss of insulation

ing them in exterior walls compromises both insulation and structural integrity. Smaller lines (up to 2 inches in diameter for most systems) are easily placed within an ICF wall without significantly compromising it, so avoiding the exterior is unnecessary in this case.

Most common for mounting small lines is to cut channels or grooves in the foam of the interior face shell and press the lines into these. This allows running the lines almost anywhere along the wall. There are several methods of making the cuts. (See Chapter 16.) This operation is unfamiliar to many electrical crews, but is easy to learn. It also removes a minor amount of insulation.

It is also possible to attach furring strips to the surface of the wall and run narrow lines between the strips. This is familiar to some crews from finishing basements or commercial concrete structures. It bears the added expense of the furring and leaves space with a depth only equal to the thickness of the strips. In some cases this will not be sufficiently deep for mounting electrical or plumbing lines to meet local code requirements. For all of these reasons, furring strips are used only occasionally. They can of course be used in conjunction with

Attachments 169

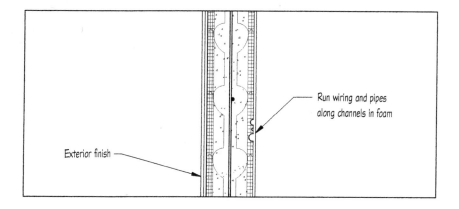

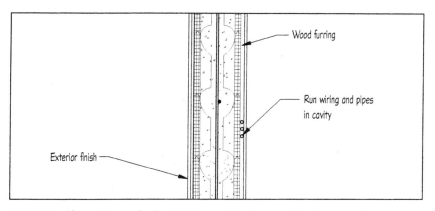

Figure 7-10 Alternative methods of placing utility lines outside the concrete.

cutting channels in the foam to allow mounting of larger-diameter lines and electrical boxes.

Some crews place pipes and conduit (for electrical wiring) in the ICF core before the pour. The result is permanent channels for the utility lines inside the wall. They are believed to be relatively secure in this position, and almost any size of line might be run here. This method requires advance planning and, usually, visits by the plumber and the electrician before the pour. Particularly large pipes or conduit may create lines of weakness in the wall. Structural analysis would be necessary to determine this precisely. Some metal pipes are not appropriate for this method because their materials can react with the

170 Design

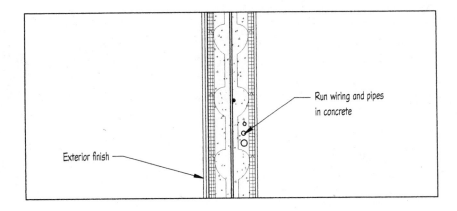

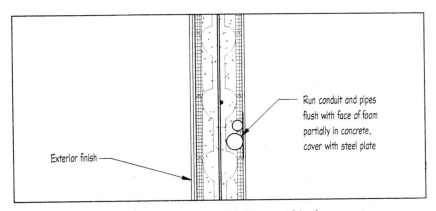

Figure 7-11 Alternative methods of placing utility lines within the concrete.

concrete. Lastly, getting access to the lines after construction can be difficult.

A variation on placing lines inside the wall is to cut a slot out of the foam face shell of the wall for the line, then place it in the foam *and* into the core. In this way the intrusion of a large line into the core is reduced. But the general pluses and minuses of this method are similar to those for complete in-concrete mounting. Some codes require, and conservative practice dictates, covering the exposed lines with steel plates to guard against later accidental penetration. Even if this is not done, it is usually necessary to cover the cut formwork temporarily to seal and reinforce it during the pour.

TABLE 7-6 Interior Finish Alternatives

Interior finish	Limitations	Advantages	Disadvantages
Unfurred wallboard	None	Familiarity	(Sometimes) special provisions necessary
Furred wallboard	None	Provides options for utility lines	Cost, time
Direct plaster	None	Material savings	Uncertain durability, (sometimes) special provisions necessary

Interior Finish

The three commonly used methods of finishing the interior surface of ICF walls are listed in Table 7-6. Figure 7-12 contains details.

Most common is to attach wallboard and finish it conventionally. This is straightforward on systems that have fastening surfaces. Plastic, steel, and wood surfaces all accept wallboard screws. Crews must plan and do a little more work to attach to systems without fastening surfaces, but this is not unduly difficult or time-consuming.

Occasionally the wall is furred first. This step can also facilitate the running of utility lines (discussed earlier in this chapter). However, it increases cost and construction time.

A few contractors plaster the formwork directly. This is common practice over the systems made of foam-cement composites. Their stiffness provides backing for plaster, and their very rough surface promises to provide good adhesion. Requiring no wallboard, it is also economical. Direct plastering has also been done on the pure foams, and this is common in other parts of the world. The foams with a rough surface promise good adhesion. Their flex has made some concerned about durability, however. Some crews first cover the wall with steel or fiberglass mesh (as is done for exterior stucco). This should strengthen the plaster, but diminishes the cost savings.

Exterior Finish

Above grade, ICFs can take any exterior finish popular on frame, including:

Stuccos

Clapboard and synthetic clapboard (e.g., fiber-cement boards) siding

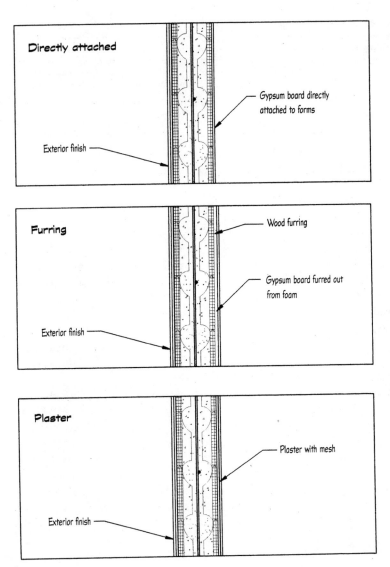

Figure 7-12 Alternative interior finishes.

Vinyl, aluminum, and steel siding
Hardboard and other panels
Masonry veneers
Shingles and shakes

Considerations of which finish to use are the same as when build-

Figure C-1 Contemporary home incorporating several spans and projections.

Figure C-2 Detail of a projection.

Figure C-3 Cost-effective bungalow.

Figure C-4 Contemporary home finished with vertical tongue-and-groove siding. *(Reddi-Form Inc.)*

Figure C-5 Ocean-front home sided with cedar shingles.

Figure C-6 ICF mansion. *(American ConForm Industries Inc.)*

Figure C-7 Brick-veneered contemporary home. *(ENERGY LOCK Inc.)*

Figure C-8 Southwestern style house with stucco finish. *(American Polysteel Forms.)*

Figure C-9 Irregular openings and trim of foam covered with stucco in a Georgian home.

Figure C-10 Use of glass blocks and window roll screens to achieve a European style.

Figure C-11 Stone veneer on an ICF addition (right) to a small round stone structure (left).

Figure C-12 ICF units visible under the eaves of a stone veneer home.

Figure C-13 Use of curved walls in an ICF California home. *(American ConForm Industries Inc.)*

Figure C-14 Vinyl siding and stucco on an Oregon house.

Figure C-15 Southern California style home.

Figure C-16 Economical ranch house.

ing with frame. Aesthetics, durability, and (with a few exceptions) cost are the same regardless of whether they are placed over ICFs or frame. Stucco is the first exception to the rule that costs are equal. Because no backing membrane need be applied to an ICF wall, stuccoing is approximately $0.50 per square foot less than on a frame wall in most areas. Note that this is true regardless of the type of stucco one chooses: whether a traditional (also called portland cement or PC) stucco, which consists primarily of cement and sand, or a polymer-based or thin-coat (also called PB) stucco, which includes about half acrylic polymer.

The second exception to the cost rule applies to sidings that must be nailed or screwed to the wall. These sometimes require exterior furring strips to be attached first, at an added cost of about $0.30 per square foot. Table 7-7 summarizes the situation.

Stucco never involves furring strips. Clapboard must generally be nailed. Nailing directly to the fastening surfaces is possible when they are wood or plastic. But steel does not grip nails well. Therefore furring strips must first be screwed to steel fastening surfaces. And systems without fastening surfaces of course require furring before clapboard. Figure 7-13 provides a detail for furring and clapboard.

Vinyl, aluminum, steel, and panels can generally be attached with screws. This is possible with all fastening surfaces, so only walls with no fastening surfaces of any type require furring. The same furring details used for clapboard are adequate for these finishes.

Masonry is connected to the wall with periodic steel ties. As they can be screwed to fastening surfaces or preinstalled in the wall and locked into the concrete, no system requires furring for a masonry finish. Figure 7-14 provides a detail.

Shingles and shakes must generally be nailed, and at intervals closer together than the fastening surfaces of any ICF. Thus furring is virtually always required. For these finishes the furring must be horizontal and spaced to result in the desired shingle/shake exposure, as

TABLE 7-7 Furring Requirements of Above-Grade Exterior Finishes

Finishes	ICF fastening surface material			
	Wood	Plastic	Steel	None
Stucco	No	No	No	No
Clapboard, synthetic clapboard	No	No	Yes	Yes
Vinyl, aluminum, steel	No	No	No	Yes
Hardboard, panels	No	No	No	Yes
Masonry veneer	No	No	No	No
Shingles, shakes	Yes	Yes	Yes	Yes

174 Design

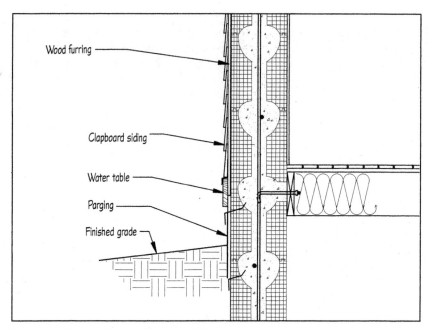

Figure 7-13 Furred clapboard exterior finish.

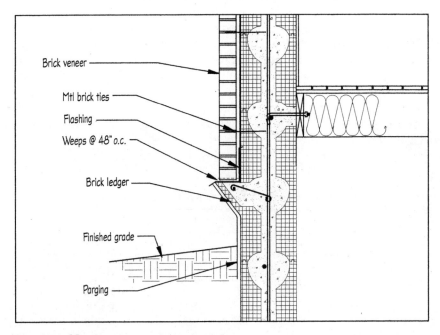

Figure 7-14 Masonry veneer exterior finish.

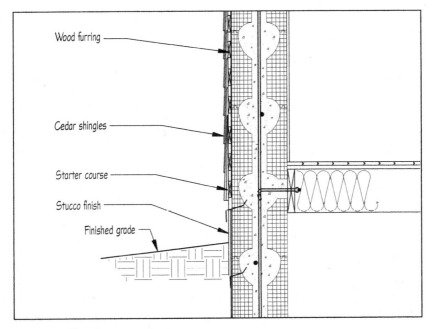

Figure 7-15 Shingle exterior finish.

illustrated in Figure 7-15. An alternative that saves labor and may be less expensive is simply to sheath the entire exterior with thin plywood before attaching shingles.

Even when it is not required, furring is sometimes beneficial. Table 7-8 lists the considerations. When finishing with clapboard, vinyl, aluminum, steel, or panels, furring creates an air space through which water can drain to the ground and any accumulated moisture can evaporate. Whether on an ICF or on frame, this adds a measure of protection against moisture infiltration and slows deterioration of wood-based sidings. It is not usually done on either type of wall, how-

TABLE 7-8 Siding Mounting Alternatives

Siding mounting	Limitations	Advantages	Disadvantages
Flush	Systems with fastening surfaces	Low cost, simplicity of installation	Lack of separate moisture evacuation
Furred	None	Potential moisture evacuation	Cost, additional installation steps

TABLE 7-9 Finish Alternatives for Exposed Foundation Walls

Finish	Limitations	Advantages	Disadvantages
Stucco	None	Impact resistance, appearance	Time, cost
Dampproofer or waterproofer	None	Speed, low cost	Low impact resistance, poor appearance

ever, because it adds time and cost, and the problems it addresses are seldom severe.

Over the exposed portion of an ICF foundation wall, stucco is by far the most popular finish. This is largely because it is highly resistant to moisture, such as frequently collects at or below ground level. It can also be attractive. Some contractors finish the exposed foam with the dampproofer or waterproofer they used below grade. This is fast and inexpensive. However, it leaves the foam vulnerable to physical damage (see Chapter 1). Table 7-9 lists the key features of these two alternatives.

References

American Society of Civil Engineers, 1994
    Standard for the Structural Design of Composite Slabs. New York: American Society of Civil Engineers.

Canadian Portland Cement Association, 1995
    Concrete Housing Handbook. Ottawa, Ont.: Canadian Portland Cement Association.

Canadian Sheet Steel Building Institute, 1988
    Criteria for the Design of Composite Slabs, Report CSSBI S3-88, Canadian Sheet Steel Building Institute, Willowdale, Ont., November.

Heagler, Luttrell, and Easterling, 1991
    Heagler, Richard B., Larry D. Luttrell, and W. Samuel Easterling, *Composite Deck Design Handbook.* Canton, OH: Steel Deck Institute.

Chapter

8

Cost Estimation

Four types of cost estimation are useful to the planning and design of a project: preliminary, sensitivity, itemized, and bid.

Preliminary estimation uses a simple multiple of square footage. It is useful for verifying that ICFs are appropriate for the project and setting initial budget targets. *Sensitivity estimation* adjusts the preliminary figure up or down for specific features of the proposed design. It produces a slightly better estimate of the total cost and shows how cost would be affected by design changes. *Itemized estimation* involves estimating the required quantity of all components and labor hours, then pricing them to arrive at a total cost. It produces the most detailed and, hopefully, most accurate independent estimate and allows for some more detailed sensitivity analysis. A *bid estimate* consists simply of sending project plans out to contractors to get their price quotes.

A preliminary estimate based on square footage can be done as described under "Cost" in Chapter 1. At the other extreme, receiving bids is normally a part of every project, and we do not cover it here.

It is not absolutely necessary to perform either of the two independent estimates that provide cost breakdowns (sensitivity and itemized). The second can be especially time-consuming. The required effort may even outweigh the insight gained. But they are worth considering. The greater detail they provide helps the designer think through the fine points of construction, quantify design tradeoffs, and improve the insight one brings to bear in negotiating with and controlling contractors. The exercise of detailed independent estimation may be particularly useful to designers inexperienced with ICFs.

Sensitivity Estimation

As a baseline estimate of total cost, one can use the rules given under "Cost" in Chapter 1. A reasonable approach is to start with the center of the range given: $2.50 more than frame per gross square foot of exterior wall area for an experienced ICF wall crew.

Although countless factors can affect the cost of construction on a particular project, some have a relatively large and predictable impact. Table 8-1 quantifies these with average figures derived from our 1995–1996 interviews. Note, however, that these numbers can vary sharply by region. They will also vary over time if ICF costs fall as expected.

Lumber price

The cost of frame construction is more affected by lumber prices since ICFs use much less wood. Our cost estimates came from early 1996, when lumber traded on commodities exchanges at approximately $400 per thousand board feet. Table 8-1 presents the impact of a fall to $300 (approximate 10-year low) or a rise to $550 (approximate 10-year high) on the difference between ICF and frame costs.

Concrete price

Concrete prices are more stable over time than those of lumber. However, they do vary by region from the $54 per cubic yard assumed in our original estimates. If they are higher or lower in your area, the extra cost of using an ICF rises or falls, respectively.

Formwork price

Prices for all the components that go into ICF formwork were about $2.50 per square foot in early 1996, delivered. But the simplest systems could be purchased in some regions of the country for $2.00 or less. Conversely, buying the most complex in other regions increased cost to $3.00 or more. These two extremes would decrease or increase the cost of ICFs (and their cost relative to frame) by $0.50. Note, however, that this does not include secondary effects. More complex (and more expensive) ICF systems generally have extra features designed to, among other things, decrease labor and other materials costs. Thus the net impact on installed cost of using a more expensive but more feature-laden ICF is uncertain until the characteristics of the structure are taken into account.

Exterior finish

Finishing the exterior with stucco is less expensive on an ICF wall. The ICF's continuous foam surface is usually a sufficient substrate, while frame requires a sheathing. Conversely, finishing with a siding that must be fastened to the wall more frequently than every stud

TABLE 8-1 Cost Impacts of Important Construction Variables

Construction variable	Estimated cost increase (decrease) ($/sq ft of gross wall area)		
	Cost of using ICFs	Cost of using wood frame	Cost premium of using ICFs vs. wood frame
Lumber prices[a]:			
$300/1000 bd ft	(0.04)	(0.16)	0.12
$550/1000 bd ft	0.06	0.24	0.18
Concrete prices[b]:			
$35/cu yd	(0.16)	0	(0.16)
$65/cu yd	0.13	0	0.13
ICF form prices[c]:			
$2.00/sq ft	(0.50)	0	(0.50)
$3.00/sq ft	0.50	0	0.50
Exterior finish:			
Stucco	2.00–4.00	2.50–4.50	(0.50)
Shingle or shake	3.80	3.50	0.30
Curved walls[d]	1.00	1.40	(0.40)
Large projections or spans	To be estimated by situation		
Reinforcement for disaster:			
High wind	0	0.50	(0.50)
Earthquake	0.20	0.05	0.15
Accurate sizing of HVAC	(0.75)	0	(0.75)
Addition of air exchange[e]:			
Intake	0.15	0	0.15
Air-to-air exchanger	0.75	0	0.75

[a]Assumes 1.6 board feet of lumber per square foot of frame wall, including dimension lumber, sheathing, waste, and other uses; assumes 0.4 board foot of lumber per square foot of ICF wall, consisting mostly of buck lumber and waste from lumber used in temporary bracing.

[b]Assumes 0.012 cubic yard of concrete per square foot of ICF wall, which is an approximate average figure. Actual concrete content varies significantly across the various systems. (See "Concrete Content" in Chapter 5.)

[c]Estimates do not include potential cost savings from added features of more expensive systems.

[d]Cost changes apply to curved sections of wall only.

[e]This is only an incremental cost to ICFs if one elects to include air exchange in the ICF house and no air exchange would have been included in a frame version of the house.

width (such as individual wooden shingles or shakes) is usually more expensive on ICFs because added strapping is required to form a nailing surface. On frame the plywood sheathing is sufficient as the backer. Other common sidings (vinyl; lapped board; panels of hardboard, composite material, or shingles) may be installed without strapping, depending on the system used. Consult "Exterior Finish" in Chapter 7 for details.

Curved walls

Most irregular wall features (90° angles, irregular angles, nonrectangular openings) add about as much cost to an ICF wall as to a frame wall, so they are not important for comparative purposes. However, according to reports from contractors, obtaining a true curve on a wall averages about $0.30 per square foot less with an ICF. The logistics are described under "Curved Walls" in Chapter 14. Note that this cost differential applies to the curved portion of the wall only.

Spans and projections

Building a house with exterior walls that do not align vertically can result in a greater incremental expense for ICFs. Building unsupported walls of ICFs requires original engineering and temporary support while the concrete is being placed. "Spans and Projections" in Chapter 6 and "Design Extremes" in Chapter 9 discuss the mechanics in more detail. The extra cost is highly situation-specific and must be estimated separately in each case. It is often avoided in practice by constructing the unsupported second-story wall of conventional frame. However, this loses the benefits of ICF walls in those areas.

Disaster provisions

Meeting the strictest wind codes in the United States (generally, in coastal states of the Southeast and around the Gulf of Mexico) rarely requires any change to ICF walls. They have an inherent high-wind resistance by virtue of their great weight. (See "Wind" in Chapter 9.) However, builders in Florida, where codes have recently become more stringent, report that there is an incremental cost to meeting high-wind requirements with wood frame.

Conversely, making a wall sufficiently resistant to earthquake for the strictest seismic code provisions (mostly in the Far West) is more expensive when using an ICF. Few extra measures are necessary on frame houses. But conservative design of the ICF house calls for increasing the amount of steel reinforcing bar used, compared with average levels.

HVAC sizing

As described under "Energy Efficiency" in Chapter 1 and in Chapter 11, the greater energy efficiency of houses built with ICF walls makes it possible to install a much smaller HVAC system. Generally speaking, a system of one-half to three-quarters the capacity required for a frame version of the house will be more than sufficient. Our figures for an average home suggest this could amount to $1000 to $2000 in reduced initial cost.

But these numbers are dependent on retaining an appropriately qualified HVAC contractor. Since ICF houses are more highly insulated and tighter than the typical house, contractors inexperienced with highly energy-efficient buildings might tend to specify heating and cooling equipment as they would for a frame home of the same size. This misses the opportunity to realize the savings.

Air change

Air exchange is increasingly recommended for all new construction, regardless of the wall system used. In Canada it is usually mandatory. Thus it may not affect the difference in costs between ICF and frame homes. However, some buyers who would not normally include it might choose to do so on an ICF home because of the wall's greater air tightness. (See "Air Infiltration" in Chapter 11.) In this instance, the equipment is an incremental expense to building with ICFs. For an average U.S. home a simple fresh-air intake vent will typically add $100 to $200 to the initial cost, an electrical air-to-air heat exchanger about $1000 to $2000.

Itemized Estimation

Our recommended procedure for itemized cost estimation is as follows. (1) List all significant materials and labor and estimate the quantity of each of these. (2) Multiply each item by an appropriate waste factor. (3) Find or estimate the unit price of each item. (4) Multiply the quantity of each item by its price to get a cost for the items. (5) Add these costs to arrive at the total project cost.

Some ICF manufacturers perform takeoffs as a service to users of their system. You may therefore be able to have some or all of the steps of itemized cost estimation performed for you. Check with your manufacturer for details.

Table 8-2 provides a spreadsheet to do the complete job yourself. You may wish to make copies of this or put it on a computer spreadsheet. The following sections discuss each step.

TABLE 8-2 Project Cost Estimation Spreadsheet

(A) Item	(B) Units	(C) Quantity required in place		(D) Waste factor		(E) Total quantity required		(F) Unit price		(G) Total item cost
1. Gross exterior wall area	Square feet									
2. Net exterior ICF wall area	Square feet									
3. Specialty units										
Type _____	Number	_____	× 1._____	=	_____	×	_____	=		
Type _____	Number	_____	× 1._____	=	_____	×	_____	=		
Type _____	Number	_____	× 1._____	=	_____	×	_____	=		
4. Standard ICF wall area	Square feet									
5. ICF units										
Panels	Number	_____	× 1._____	=	_____	×	_____	=		
Planks	Number	_____	× 1._____	=	_____	×	_____	=		
Blocks	Number	_____	× 1._____	=	_____	×	_____	=		
Foam faces	Number	_____	× 1._____	=	_____	×	_____	=		
Corner faces	Number	_____	× 1._____	=	_____	×	_____	=		
Sheet foam	Number	_____	× 1._____	=	_____	×	_____	=		
Standard ties	Number	_____	× 1._____	=	_____	×	_____	=		
Half-ties	Number	_____	× 1._____	=	_____	×	_____	=		
Corner ties	Number	_____	× 1._____	=	_____	×	_____	=		
Half-corner ties	Number	_____	× 1._____	=	_____	×	_____	=		

Fastening surface	Lineal feet	_____	× 1. _____	=	_____
Connecting channel	Lineal feet	_____	× 1. _____	=	_____
Other _____	Number	_____	× 1. _____	=	_____
Other _____	Number	_____	× 1. _____	=	_____
6. Concrete	Cubic yards	_____	× 1. _____	=	_____
7. Steel reinforcing bar					
4' #3	Number	_____	× 1. _____	=	_____
8' #3	Number	_____	× 1. _____	=	_____
10' #3	Number	_____	× 1. _____	=	_____
4' #4	Number	_____	× 1. _____	=	_____
8' #4	Number	_____	× 1. _____	=	_____
10' #4	Number	_____	× 1. _____	=	_____
4' #5	Number	_____	× 1. _____	=	_____
8' #5	Number	_____	× 1. _____	=	_____
10' #5	Number	_____	× 1. _____	=	_____
Other _____	Number	_____	× 1. _____	=	_____
Other _____	Number	_____	× 1. _____	=	_____
Other _____	Number	_____	× 1. _____	=	_____
Other _____	Number	_____	× 1. _____	=	_____
Other _____	Number	_____	× 1. _____	=	_____
8. Equipment rental					
Concrete pump	Half-days	_____	× 1.0 _____	=	_____
Lift	Days	_____	× 1.0 _____	=	_____

TABLE 8-2 Project Cost Estimation Spreadsheet (*Continued*)

(A) Item	(B) Units	(C) Quantity required in place	×	(D) Waste factor	=	(E) Total quantity required	×	(F) Unit price	=	(G) Total item cost
9. Bucks (mat'l _____)										
Window jambs and lintels										
(Dimensions _____)	Lineal feet	_____	×	1._____	=	_____	×	_____	=	_____
Window sills										
(Dimensions _____)	Lineal feet	_____	×	1._____	=	_____	×	_____	=	_____
Door jambs and lintels										
(Dimensions _____)	Lineal feet	_____	×	1._____	=	_____	×	_____	=	_____
10. Plates (mat'l _____)										
(Dimensions _____)	Lineal feet	_____	×	1._____	=	_____	×	_____	=	_____
11. Adhesive (connect units)										
Type _____	Tubes/cans/rolls	_____	×	1._____	=	_____	×	_____	=	_____
12. Adhesive foam (fill and seal)	Cans	_____	×	1._____	=	_____	×	_____	=	_____
13. Fasteners										
Type _____	Number	_____	×	1._____	=	_____	×	_____	=	_____
Type _____	Number	_____	×	1._____	=	_____	×	_____	=	_____
Type _____	Number	_____	×	1._____	=	_____	×	_____	=	_____
Type _____	Number	_____	×	1._____	=	_____	×	_____	=	_____
Type _____	Number	_____	×	1._____	=	_____	×	_____	=	_____
Type _____	Number	_____	×	1._____	=	_____	×	_____	=	_____
Type _____	Number	_____	×	1._____	=	_____	×	_____	=	_____

14. Labor

Type _____ Worker-hours _____ × 1.0 _____ = _____

Type _____ Worker-hours _____ × 1.0 _____ = _____

Type _____ Worker-hours _____ × 1.0 _____ = _____

Type _____ Worker-hours _____ × 1.0 _____ = _____

15. HVAC savings, or extra cost

Savings from smaller equipment _____

Cost of additional air exchange equipment (if any) _____

16. Total project cost _____

Step one: Itemization

First determine the items of material and labor that will be used by the project, and record the estimated amount of each on the form in Table 8-2.

ICF units. If your system consists of preassembled units, you can simply record the numbers of each type of unit required. The table provides in line 3 space for up to three types of specialty units and in line 5 spaces for standard units (such as panels, planks, or blocks, as appropriate). When using field-assembled units it will be necessary to estimate the number of each component of the units. Spaces for all major components are also provided in Table 8-2.

To estimate the number of units necessary for any preassembled system:

1. Calculate the gross exterior wall area. Include all openings (windows and doors), any wall segments to be constructed of frame (such as a projection), and, if they are to be constructed of ICFs, the foundation walls. Record this as gross exterior wall area in item 1.

2. From the gross exterior wall area subtract the area of all wall segments to be constructed of frame and all large openings (anything bigger than 4 square feet). Call this the net exterior ICF wall area and enter it in item 2. Note that you will have to estimate the cost of any wall sections built of frame separately.

3. If you will be using specialty units (preformed or precut corners, brick-ledge units, and so on), count the number that you will need of each. Record these numbers in item 3. Then find in the manufacturer's literature the area of the exterior surface of these units, multiply that by the number of units you will need, and subtract the resulting total area of special units from the net exterior ICF wall area you calculated in step 2. Call this the standard ICF wall area and enter it in item 4.

4. Determine the exterior surface area of each standard unit. If this is not readily available otherwise, you can read it from Table 8-3 (for commonly used sizes) or calculate it by multiplying the height and length dimensions of your system's standard units from Table 2-1 (for almost any standard unit).

5. Divide the standard ICF wall area by the area of a standard unit to get the total number of standard units required. Enter this in item 5.

For systems involving field assembly, calculate the number of units required as in steps 1 to 5. But before making entries for units into Table 8-2, break this down into the number of components of each type

TABLE 8-3   Cost Estimation Data for ICF Systems

System	Total width (inches)	Wall area (sq ft)	Concrete thickness (inches)*	Concrete (cu yd/sq ft of net ICF wall area)
Flat panel systems:				
Lite-Form (preassembled)	12	32	8	0.025
R-FORMS	8	32	4	0.0125
	10	32	6	0.0188
	12	32	8	0.025
	14	32	10	0.0313
Grid panel systems:				
ENER-GRID	8	12.5	4.5	0.0064
	8	25	4.5	0.0064
	10	12.5	6.2	0.0096
	10	25	6.2	0.0096
	12	12.5	6.2	0.0096
	12	25	6.2	0.0096
RASTRA	8½	12.5	4	0.0064
	8½	25	4	0.0064
	10	12.5	6	0.0096
	10	25	6	0.0096
	12	12.5	6	0.0096
	12	25	6	0.0096
	14	12.5	6	0.0096
	14	25	6	0.0096
Post-and-beam panel systems:				
Amhome	9⅜	32	5½	0.0044†
ThermoFormed	8	32	5	0.0063†
Flat plank systems:				
Diamond Snap-Form	8	8	4	0.0125
	10	8	6	0.0188
	12	8	8	0.025
	14	8	10	0.0313
Lite-Form (nonassembled)	12	5.33	8	0.025
Polycrete	10⅝	8	5⅝	0.0176
	11	8	6	0.0188
	12⅝	8	7⅝	0.0239
	13	8	8	0.025

*Thickness of the concrete at its thickest point.
†Amounts of concrete per unit vary with post-and-beam systems, where only select cavities are filled with concrete. The number of cavities varies with the required strength of the wall and the number of openings. The data here assume cavities filled every 48 (Amhome) or 24 (ENERGYLOCK, Featherlite, KEEVA) inches on center vertically. If more frequent cavities are filled, the amount of concrete used increases proportionately.

TABLE 8-3 Cost Estimation Data for ICF Systems (*Continued*)

System	Total width (inches)	Wall area (sq ft)	Concrete thickness (inches)*	Concrete (cu yd/sq ft of net ICF wall area)
Polycrete (*Cont.*)	14⅝	8	9⅝	0.0301
	15	8	10	0.0313
	16⅝	8	11⅝	0.0364
Quad-Lock	8⅛	4	3⅝	0.0113
	10⅛	4	5⅝	0.0176
	12⅛	4	7⅝	0.0238
	14⅛	4	9⅝	0.0294
Flat block systems:				
Blue Maxx	11½	5.58	6¼	0.0199
	12⅝	5.58	8	0.0258
Fold-Form	8	4	4	0.0125
	10	4	6	0.0188
	12	4	8	0.025
GREEN-BLOCK	9.88 (250 mm)	2.69	5¾	0.0178
SmartBlock VWF	8	3.33	3¾	0.0116
	10	3.33	5¾	0.0177
	12	3.33	7¾	0.0239
	14	3.33	9¾	0.0301
Grid block systems:				
I.C.E. Block	9¼	5⅓	6⅜	0.0139
	11	5⅓	8	0.0188
Insulform	9.6	5⅓	5¾	0.0156
Modu-Lock	10	2.77	6½	0.0186
Polysteel	9¼	5⅓	6⅜	0.0139
	11	5⅓	8	0.0188
Reddi-Form	9⅝	4	6	0.0125
REWARD	9¼	5⅓	6⅜	0.0139
	11	5⅓	8	0.0188
SmartBlock SF 10	10	2.69	6½	0.0186
Therm-O-Wall	9¼	5⅓	6⅜	0.0139
	11	5⅓	8	0.0188
Post-and-beam block systems:				
ENERGY LOCK	8	2⅔	5	0.0067†
Featherlite	8	8/9	5	0.0067†
KEEVA	8	2⅔	5	0.0056†

*Thickness of the concrete at its thickest point.
†Amounts of concrete per unit vary with post-and-beam systems, where only select cavities are filled with concrete. The number of cavities varies with the required strength of the wall and the number of openings. The data here assume cavities filled every 48 (Amhome) or 24 (ENERGYLOCK, Featherlite, KEEVA, ThermoFormed) inches on center vertically. If more frequent cavities are filled, the amount of concrete used increases proportionately.

required to assemble the units. That information should be in the manufacturer's documentation or available from the manufacturer directly.

Concrete. To estimate the amount of concrete necessary to fill the formwork:

1. Find the cubic yards per square foot for the system you are using from Table 8-3. This is the yards of concrete needed to fill one square foot of formwork.
2. Multiply the cubic yards per square foot by the net exterior ICF wall area previously calculated. Enter the result in item 6 of Table 8-2.

Note that the cubic yards per square foot will vary with post-and-beam systems. The stronger the wall needs to be, the more cavities will be filled with concrete, and the greater the concrete per square foot. The data in Table 8-3 assume concrete posts every 24 inches vertically on center. For more frequent concrete, the cubic yards per square foot will need to be increased proportionately. Consult the manufacturer's literature or the manufacturer.

Rebar. The size and amount of rebar needed will vary widely with such load factors as wall height, number of stories, local wind and seismic conditions, and number and width of openings. The most reliable method of determining these data is to tally them from the complete structural design of the house. If the structural design is not yet done, one can still estimate the rebar requirements.

A conservative rule of thumb is that a house in a low- or moderate-risk seismic area will be engineered with 0.9 lineal foot of #4 ($\frac{4}{8}$-inch-diameter) bar and 0.1 lineal foot of #5 ($\frac{5}{8}$-inch) bar per gross square foot of wall area. For high seismic areas (zone 4 or seismic performance category E), increase these figures by 50 percent. Note, however, that *these are rough generalizations of average reinforcement requirements intended only for preliminary estimation of approximate cost. Determination of actual reinforcement requirements depends on engineering analysis and will differ from these figures for any particular building.* Enter the estimates in Table 8-2, item 7.

Equipment. Typical concrete equipment requirements are one half-day rental of a concrete pump for each above-grade story of the house. (See Chapter 15 for more detail on concrete placement.) Add one more half-day for tall (greater than 4-foot) gable ends, gambrel ends, or parapet walls that will be built of ICFs. Enter the number of half-days in Table 8-2, item 8.

If you are using a system made of foam-cement composite, plan to rent lifting equipment (a small forklift truck or manually operated equivalent) for about 4 days per story. The time will be longer with large houses, and it will decline as the wall crew gains experience and works faster. Enter the total number of days also in item 8.

Buck material. Dimension lumber is often used to build bucks for openings. The designer may instead opt to mount windows and doors with a channel buck. (See "Window and Door Mounting" in Chapter 6 for more detail on channels.) Since the bucks will be in contact with concrete, some codes and designers may prefer or require that if they are lumber they be moisture-resistant. Thus they may be all pressure-treated, the buck jambs and lintels may be kiln-dried lumber wrapped with a water-resistant membrane, and the buck sills may be pressure-treated, or all buck and plate lumber may be kiln-dried with a membrane.

First specify the materials and dimensions of each item. In the blank provided in item 9, column (A), write pressure-treated, kiln-dried, or plastic or steel channel, as appropriate.

Next decide on the dimensions of the lumber (if it is used) and write these in the blanks provided. For jambs and lintels of protruding bucks, you will need the width of the lumber that spans the thickness of the entire ICF wall. The wall thickness for ICFs is available in Table 8-3, listed as "total width." If the wall thickness does not match a standard lumber width, you will need to buy a width just greater than the wall thickness and have it trimmed to size. For recessed bucks, the jamb and lintel width should be equal to the concrete thickness (Table 8-3) instead.

The sill lumber for bucks should be about $1\frac{1}{2}$ inches narrower than half the thickness of the jambs and lintels. This generally leads to the use of 2×3 or 2×4. The narrower lumber allows the sill to be constructed of two narrow pieces separated by a gap of 3 to 4 inches. Write this in item 9 after sill dimensions.

To estimate the lineal feet of lumber or channel for the bucks [column (C)]:

1. Measure the dimensions (width and height) of each opening.
2. Add the lengths of all jambs and lintels to get their total length, and the lengths of all sills to get a separate total.
3. Enter the total length of the jambs and lintels in item 9.
4. Double the total sill length calculated in step 2 and enter that number also in item 9.

Plates. If the house will use steel straps for attaching roof members (see "Roofs" in Chapter 7), there will be no plates and you may skip this item. Otherwise it is necessary to enter the lumber required for the top plates.

First decide what material (pressure-treated or kiln-dried lumber) will be used and write this into the space provided in item 10, column (A). Also write in the dimensions of the plates. The width of protruding plates may be less than the wall thickness, but never less than a nominal 4 inches. The width of recessed plates should match the concrete width of the system used, or be slightly wider for trimming to the concrete width.

To estimate the lineal feet of top plates, simply take the length of the exterior perimeter of the top story of the building. In the case of uneven roof lines (such as gable or gambrel ends), take into account the greater distance the plates must span along the inclines. Write this total in item 10, column (C).

Adhesive. Almost every system requires some form of adhesive (glue or tape) to hold together at least some joints or breaks in the formwork.

Some manufacturers recommend gluing every joint between units to prevent them from separating during the pour. Other manufacturers design their interconnects to hold by friction without glue, and even where extra steps need to be taken to hold units together, there are several methods available. (See "Interconnection" in Chapter 14.) If you choose to glue all joints, for an average-size house expect to use about one can of industrial foam adhesive or six tubes of conventional construction adhesive per story. Enter the appropriate number in item 11, column (C).

For filling and sealing holes and gaps you will also need some adhesive foam. (Construction adhesive or tape are not appropriate here.) One-tenth of a can per story should be enough (item 12).

Fasteners. The fasteners required vary widely from project to project, but usually their total cost is about the same as for frame walls of the same size. Some of them are more expensive than nails, but you will use fewer. List the heavy structural fasteners (J-bolts, metal straps) you think you will need to make connections to floor decks, roof plates or members, and load-bearing walls. For all other connectors it is reasonable to add about the same allowance you would make for connectors in the exterior walls of a comparable frame house (item 13).

Labor. Labor is the least reliably estimated of all important resources. The amount required depends heavily on the skill level of the workers, the extent of their experience with ICFs, and the complexity of the house.

A typical crew assigned to ICF construction consists of one skilled worker (the equivalent of a lead framing or forms carpenter), one semiskilled worker (the equivalent of the second person on a framing or forms crew), and two unskilled laborers (the equivalent of carpenters' helpers). Assume that such a crew builds the exterior ICF walls of a house of average complexity (all right-angle corners, 15 percent of above-grade gross exterior wall area in openings, all openings rectangular except for one or two curved or irregular). If they have *no* experience at building with ICFs but have received a basic introduction and reviewed the manufacturer's literature, they might be expected, as a crew, to complete 50 square feet of wall per hour. Thus a plausible estimation procedure is as follows:

1. Take the gross exterior wall area from item 1.
2. Divide by 50 to get the total number of crew-hours to build the walls. Enter the result in item 14 once for each of the four workers.

If the same crew has built at least three complete ICF houses before, they might be expected to accelerate to 75 square feet per hour. So you can use the same estimation procedure but divide by 75 in step 2 instead of by 50.

Highly experienced crews report even higher production rates. When dealing with one of these, use an average rate from their past few jobs.

HVAC. If you wish to take account of the HVAC system in estimating the cost of ICF exterior walls, a conservative assumption is that all heating and cooling equipment will be three-quarters the size it would be if the house were built of frame. Note, however, that the savings are dependent on retaining an HVAC contractor capable of accurately sizing equipment for a superinsulated house. Enter the savings in item 15. As discussed earlier in this chapter, if the designer or the occupants would not install air exchange in a frame house, but will in the ICF house, that will be an incremental cost of using the ICF system, which should be included in the estimates (see "Energy Efficiency" in Chapter 1) in item 15.

Step two: Estimating waste factors

Use column (D) in Table 8-2 to write estimated waste factors for the materials used in the project. There is no waste factor for labor; the time requirements of any rework are already included in the original time estimates. There is also no need to assume a waste factor for equipment rental.

Waste for materials with an inexperienced crew should be approximately 10 percent of in-place requirements as estimated in column (C). If using this approximate figure, write a waste factor of 1.1 for each material in column (D). A waste rate of 5 percent (waste factor 1.05) is a reasonable estimate for a crew that has built three or four houses; the most experienced ICF crews have achieved waste rates of 2 percent (waste factor 1.02).

The major exception to the typical waste factors cited in the preceding paragraph is the waste rate of lumber for bucks. Cutoffs from buck lumber are rarely reused because the labor cost of piecing together short lengths exceeds the materials cost. Therefore a waste factor of 1.1, regardless of crew experience, is probably more accurate.

To complete the accounting for waste, multiply the in-place quantity of each item as listed in column (C) by the corresponding waste factor listed in column (D). Record the resulting numbers in column (E), total quantity required.

Step three: Estimating prices

Column (F) of Table 8-2 provides space for entering unit prices for all materials and labor. The most reliable source of estimates is generally current quotes from local vendors and crews. Published price lists and estimating guides are quicker alternatives.

Bear in mind that, depending on whether you are estimating total cost to the contractor or to the end buyer, the prices may need to include a contractor's markup (overhead and profit).

Step four: Calculating total item costs

For dollar costs, multiply the total quantity required of each item in column (E) of Table 8-2 by its unit price in column (F). Enter the results in column (G).

Step five: Calculating total project costs

Sum all amounts in column (G) to get the total project cost. Enter it in item 16.

Part 4

Engineering

In this part we provide procedures for engineering the structure and energy systems of ICF homes. Note that as there exists some variation in current practice, the recommendations of some ICF manufacturers and the actual practice of some engineering professionals may differ from what is presented here.

This part and this book are not intended to substitute in any way for the recommendations of any ICF manufacturer or accepted engineering practice in general. The manufacturers' recommendations and accepted engineering practice always take precedence over any material presented here. This part and book are intended only to present general principles of ICF engineering to assist in understanding the manufacturers' documentation and the structural and energy properties of ICF buildings.

Most ICF houses have lower story heights and lighter load conditions than the typical concrete walls historically present in commercial construction. Extensive structural design procedures and aids are available for the taller, more heavily loaded walls. Here we highlight the differences in procedure for residential walls. Energy calculations also require some modification of conventional practice. Residential HVAC calculation procedures and tools rarely include provisions to account for such high R-values, low air infiltration, and high thermal mass as those present with ICFs.

Not surprisingly, methods for engineering ICF buildings are changing. New data from wall tests and more thorough applications of the codes are leading to refined, better-grounded rules and formulas for both structural and energy load calculations. Past engineering relied on consistently *selecting highly conservative assumptions because of a lack of understanding of ICF walls. As more becomes known, more exact formulas are replacing the use of the most conservative inexact formulas. Therefore, in addition to presenting current methods here, we discuss which items are likely to undergo change in the future and what that change might be.*

Chapter 9

Structural Performance

The record of concrete engineering and building performance provides a sense of the structural capabilities and limits of the ICF systems. It also tends to confirm the general rule that they exceed the capabilities of conventional wood frame and the requirements of low-rise residential construction.

Design Extremes

To date the tallest single-family homes built with all-ICF exterior walls in North America include a full basement plus three stories above grade. Commercial buildings of at least six stories have been constructed. Chapter 10 presents ICF engineering procedures that could be extended to multiple stories.

The longest lintel constructed with an ICF that we have discovered was 23 feet for a double-width garage door. Chapter 10 presents general rules for lintel design. Engineers designing lintels near this length often request use of the wider available formwork units to obtain a thicker concrete cross section. (See Table 6-2 for the concrete thicknesses available under each system.) Some have even required oversized lintels, which must be created with conventional formwork, as depicted in Figure 9-1. Form boards are set so that the placed concrete will match the full width of the ICF units. This results in more space for reinforcement and a cross section of as much as twice the thickness of the concrete inside the ICF forms. Such a lintel will have no foam insulation. This is usually unimportant in a garage, however.

Spans and projections of ICF walls have been constructed as long as 23 feet. Figure 9-2 depicts the two structural means employed to support these walls, which do not align with the walls below. One is to add enough reinforcement to the ICF wall for it to support itself

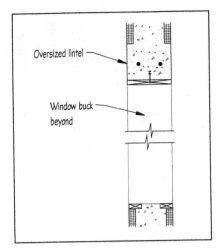

Figure 9-1 Oversized concrete lintel (built with temporary plywood formwork).

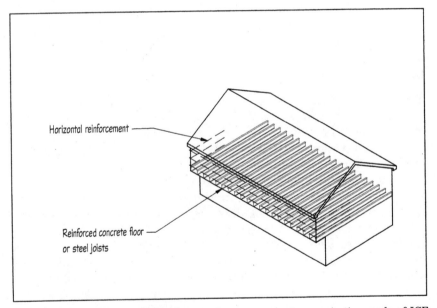

Figure 9-2 Alternative methods of supporting an upper-story projection made of ICF walls.

and all applied loads. This in effect turns the wall into a tall beam. The second is to support it on a high-strength floor deck. To date these decks have been made of reinforced concrete (see Chapter 7 for the floor systems used) or steel beams. All were custom engineered for the specific structure.

ICF homes have been built in areas subject to the highest categories of load conditions and strongest of codes: the severe ground freezes of Minnesota, hurricane-prone southern Florida, and the high seismic areas of the Southwest. To date we have no reports of structural failure of any ICF house.

Wind

There are as yet no systematic reports of the experience with ICF houses subjected to high winds. However, there is information on other types of low-rise reinforced concrete homes. This is potentially a fair indicator of what one might expect with the ICFs.

After the hurricanes Andrew (South Florida) and Iniki (Hawaii) of 1992, two study teams conducted surveys of the affected areas to determine which design and construction practices are most susceptible to wind damage. A group headed by Dr. Ronald Zollo of the University of Miami visited the areas ravaged by Andrew (Zollo, 1994), and the research arm of the National Association of Home Builders surveyed the sites of both hurricanes (NAHB Research Center, 1993).

Both studies reached similar conclusions. The most common damage was loss of all or parts of the roof in approximately 25 percent of the homes in the direct hurricane path. This loss could almost always be traced to the wind acting directly on the roof, either under the eaves or against a gable end, or to windows blowing in and the house pressurizing sufficiently to force off the roof. After loss of the roof diaphragm, houses with concrete walls rarely suffered any noteworthy further damage. In contrast, roof-damaged frame houses experienced severe wall damage a significant (but unreported) proportion of the time. The explanation given for this result is that the greater inherent weight and strength of the concrete walls are effective in resisting the high and sustained uplift and transverse forces of wind.

There exist components to reinforce the connections of frame houses and make them more resistant to wind damage. However, they add cost to the home, as discussed in Chapter 8. In addition, the University of Miami report claims that wood frame construction is less "forgiving" than concrete, in the sense that variations in material and workmanship are more likely to compromise the structure's wind resistance.

Earthquake

There exist few field data on the earthquake performance of the structures of most interest: low-rise concrete with modern reinforcement schedules. One ICF manufacturer reports that one of its homes survived a Richter 4.9 earthquake without damage (Bowman, 1995). Empirical data should accumulate as the number of ICF homes increases.

Historical conventional wisdom is that frame walls sheathed with plywood are highly earthquake-resistant, whereas concrete performs poorly if unreinforced but well if properly reinforced (Spangle, Meehan, Degenkolb, and Blair, 1987).

Anecdotal evidence from the 1993 earthquakes in Northridge, California, and Kobe, Japan, largely bear out these generalizations (*Economist,* 1995; Berger, 1994; *Los Angeles Times,* 1994). Plywood-sheathed frame walls suffered moderate damage. This damage was concentrated in midrise frame buildings and in homes apparently built with less conscientious construction practices. Tall unreinforced and lightly reinforced masonry and concrete structures apparently experienced some of the highest rates of damage, whereas more heavily reinforced concrete buildings held up significantly better.

As detailed in Chapter 10, calculations indicate that the addition of steel reinforcement can make concrete walls (including ICFs) as resistant or more resistant to earthquake than conventional frame. Code organizations are gradually increasing the structural requirements for buildings in areas rated high in seismic risk. Several ICF manufacturers have recommended schedules for steel reinforcement designed to meet the latest requirements of these higher-rated areas (zone 4 or seismic performance category E, depending on the rating scheme used). These include a minimum of as much as one #4 bar every 12 inches vertically and every 16 inches horizontally. Their calculations indicate that these schedules will satisfy the requirements for most house designs in high-earthquake zones.

The recommended reinforcement schedules reflect a preference for conservatism in designing for earthquake. It arises because the incremental cost of reinforcement is low, and the level of uncertainty is high. Seismic forces and the ability of structures to resist them are less well understood than other natural disasters. Severe earthquakes occur infrequently, so there is less opportunity for systematic surveys. Yet simulating seismic effects in the laboratory is too expensive to be done routinely. Empirical data should accumulate as the number of ICF houses and their years in place grow.

References

Berger, 1994
  Berger, Leslie, "Quake Reveals Widespread Building Defects," *Los Angeles Times*, April 21, p. 1.

Bowman, 1995
  Bowman, Lisa M., "A Foam Foundation: New Material for Building Holds Promise for Quake and Fire Victims," *Los Angeles Times*, July 11, p. 1.

Economist, 1995
  "Lessons from East and West: Both the Kobe and Northridge Quakes Provided Some Nasty Shocks for Structural Engineers," April 22.

Los Angeles Times, 1994
  "72,000 Homes Reported Affected by Earthquake," *Los Angeles Times*, March 29, p. 2.

NAHB Research Center, 1993
  "Assessment of Damage to Single-Family Homes Caused by Hurricanes Andrew and Iniki," Report prepared for U.S. Department of Housing and Urban Development and Office of Policy Development and Research, Contract HC-5911, September.

Spangle, Meehan, Degenkolb, and Blair, 1987
  Spangle, William, Richard L. Meehan, Henry J. Degenkolb, and Martha L. Blair, "Pre-Earthquake Planning for Post-Earthquake Rebuilding," Report of the Southern California Earthquake Preparedness Project, Governor's Office of Emergency Services, State of California.

Zollo, 1993
  Zollo, Ronald, "Hurricane Andrew: August 24, 1992, Structural Performance of Buildings in Dade County, Florida," Technical Report No. CEN 93-1, University of Miami, Coral Gables, FL.

Chapter 10

Structural Design

This chapter describes procedures for structural engineering with ICFs and presents sample calculations for a simple, hypothetical house. As the technical manuals of the different ICF manufacturers employ various engineering approaches, this chapter will differ from some of them in certain specifics.

The engineering procedures presented here are intended as an introduction to the structural engineering of ICF wall systems in general. *This chapter and this book are not intended to substitute in any way for the recommendations of any ICF manufacturer or accepted engineering practice in general. The manufacturers' recommendations and accepted engineering practice always take precedence over any material presented here.*

Overview

Walls built with ICFs are structurally identical to cast-in-place walls built using removable forms. The fact that the foam formwork stays in place is of little consequence since it is not relied on to carry any structural loads. The calculations required for concrete walls are generally well understood and well documented.

Almost all of the recommended procedures for U.S. users of ICF systems are based on the *Building Code Requirements for Structural Concrete* (ACI 318-95), published by the American Concrete Institute (1995). ACI 318-95 (or an earlier version) effectively serves as a model code for cast-in-place concrete construction in the United States. Its recommended procedures are adopted with limited change by all three of the major model building codes (BOCA, ICBO, SBCCI). Most local jurisdictions, in turn, adopt one of these model codes. When presenting requirements taken from the ACI 318-95, we include a reference to the relevant document and section, in the format: ACI 1.1.1.

We reference formulas from ACI 318-95 with the equation number in the format: ACI EQ 1-1.

Canadian engineering practice follows the guidelines of the Canadian Standards Association (CSA). It and ACI 318 are becoming increasingly consistent with regard to cast-in-place concrete, but there are differences in some particulars. Therefore the procedures described here will need inspection and adjustment before application in Canada.

Engineering requirements vary according to the class of ICF used. Flat walls may be designed in accordance with ACI Chapter 14 (Walls). Under some circumstances the code allows the option of instead designing in accordance with Chapter 22 (Structural Plain Concrete).

Grid walls may also be designed in accordance with either of two parts of the code: Chapter 14, Section 14.4, or Chapter 22, Section 22.6. These are the sections that apply to walls other than those with solid, rectangular cross sections.

Post-and-beam walls comprise a series of vertical and horizontal members and therefore do not fit the absolute structural definition for wall. They are better approximated as frames for engineering purposes and are not discussed in detail here. Requirements for frame design are found in ACI 318-95 Chapter 10 (Flexure and Axial Loads).

The text and numerical examples of this chapter are designed to provide information on major variants of the basic ICF. This includes identifying which steps of the engineering process apply to which classes of systems and, where useful in the numerical examples, supplying different calculations for different system classes.

Note that frequently the manufacturers' manuals, regardless of the class of system involved, include tables of the results of structural calculations for common wall conditions and configurations. These relieve the engineer of the burden of performing many routine calculations. While it would be awkward to produce tables in this chapter to cover the common variables and circumstance for all systems, the discussion should clarify the origins of the manufacturers' tabular information, as well as assist in original calculations.

Structural Checks

The engineering of ICF buildings is in principle like that of any other structural system. Calculations are done to determine loads and forces applied to each portion of the wall, and design formulas based on geometric and material properties are used to determine whether the wall has adequate strength to resist the loads. If it does not, the engineer revises until an adequate structural design is reached. Figure 10-1 depicts the various types of wall forces commonly checked in this manner.

Structural Design

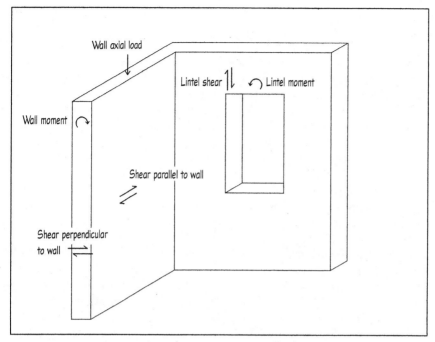

Figure 10-1 Forces commonly analyzed in ICF engineering.

In principle the engineer has several variables for influencing the strength of ICF structural members, including:

Concrete compressive strength

Concrete wall thickness

Vertical reinforcement, including:

- Size
- Spacing and number of bars
- Tensile strength
- Positioning in the wall

Horizontal reinforcement, including:

- Size
- Spacing and number of bars
- Tensile strength
- Positioning in the wall

Form cavities filled with concrete (post-and-beam systems only)

Wall height
Wall length

In practice, some of these are adjusted much less often than others. Those most often varied are the ones that are least costly and least disruptive to other aspects of the design.

The design or specified compressive strength of the concrete is rarely altered. 3000 psi is usually adequate for low-rise residential construction. The designer can specify concrete of higher compressive strengths if necessary. Concrete of up to 4000 psi is common and easily achievable at little or no additional cost.

The wall thickness is usually selected on nonstructural criteria and rarely manipulated in design. The geometry of the chosen formwork generally dictates the exact thickness or shape of the concrete inside.

Modifying the size and arrangement of the steel reinforcing bar is another method of increasing the strength of ICF members. This is accomplished by (1) increasing the diameter of the bars used or (2) decreasing the bar spacing (effectively increasing the amount of steel in the wall). Note also that in a lintel, vertical reinforcement ordinarily takes the form of stirrups.

The specified yield strength of reinforcement is typically 40,000 psi (grade 40) or 60,000 psi (grade 60). Since contractors sometimes deviate from specifications, the conservative approach is to design with grade 40 reinforcement. Although reinforcing steel, if required, is normally placed in the center of the wall, the designer may choose to position the bars closer to the face that will be in tension. This increases the wall's resistance to moment without increasing the amount or strength of any of the materials. However, it depends on the ability to predict the direction of the moments reliably.

How many and which cavities to fill with concrete is at the discretion of the designers when using post-and-beam systems. Decreasing the spacing between concrete members (effectively increasing the number of members in the wall) increases strength. It does so both directly (from the strength of the concrete) and by providing additional locations to which steel reinforcement can be added.

Reducing wall heights and lengths can reduce loads. However, engineers rarely manipulate this variable because of its interference with other aspects of design.

Reinforced Concrete Design Procedures

The provisions of ACI Chapter 14 (Walls) come under a method of engineering design informally referred to as "reinforced concrete" because of its dependence on reinforcement to achieve required strengths.

Both flat and grid systems may be engineered according to ACI Chapter 14 requirements.

Note that since some grid walls have complex cross-sectional shapes (ellipses, rectangles with rounded corners, etc.), rectangular approximations of the structural section (noted later in this discussion) are typically made for purposes of design.

Minimum reinforcement

ACI 14.3 specifies minimum amounts and spacings of both vertical and horizontal reinforcement. In some cases these prescribed limits may be excessive for a single-family home. The designer may therefore choose to waive the requirements by invoking ACI 14.2.7:

> Quantity of reinforcement and limits of thickness required by 14.3 and 14.5 shall be permitted to be waived where structural analysis shows adequate strength and stability.

This, however, makes it incumbent on the designer that the required structural analysis be performed.

Flexure and axial loads

ACI 14.2.2 allows the designer two options in designing reinforced walls: (1) walls may be designed as compression members using the strength design provisions for flexure and axial loads of ACI Chapter 10, as described in ACI 14.4, or (2) they may be designed by the empirical design method of ACI 14.5. The provisions of ACI 14.2 (General) and ACI 14.3 (Minimum Reinforcement) apply to walls designed by either method. No minimum wall thicknesses or limits on cross-sectional geometry are prescribed for walls designed as compression members (ACI 14.4).

Empirical design method

The empirical design method applies to load-bearing walls of solid rectangular cross section. This would include flat ICF walls. Minimum thicknesses are prescribed for walls designed by the method (ACI 14.5.3). Grid walls must instead be designed using the provisions in ACI 14.4 (see "Flexure and axial loads").

Another limitation of the empirical design method is that the resultant of all applied axial loads falls within the middle one-third of the wall thickness (eccentricity $e \leq h/6$). In addition, the thickness of the wall shall not be less than 1/25 of the supporting height or length of the wall, whichever is shorter, nor less than 4 inches (ACI 14.5.3.1). Basement walls designed using this method cannot be less than $7\frac{1}{2}$

TABLE 10-1 Formulas for Calculation of Flexure and Axial Load Strength of Reinforced Concrete Wall Using the Empirical Design Method

$\phi P_{nw} \geq P_u$

$$\phi P_{nw} = 0.55 \phi f_c' A_g \left[1 - \left(\frac{kl_c}{32h} \right)^2 \right]$$ (ACI EQ 14-1)

Variable	Identity	Unit
A_g	Gross area of section	in²
f_c'	Specified compressive strength of concrete	psi
h	Overall thickness of member	in
k	Effective length factor*	dimensionless
l_c	Vertical distance between supports	in
P_{nw}	Nominal axial load strength of wall	lb
P_u	Factored axial load	lb
ϕ	Strength reduction factor†	dimensionless

*Values given in ACI 14.5.2. For most residential construction applications, a value of $k = 1.0$ is appropriate if the wall is tied to the footing, floors, and roof.
†Set equal to 0.70 for flexure and axial load (ACI 9.3.2).

inches thick (ACI 14.5.3.2). Note that in addition to any eccentric axial loads, the effect of any lateral loads on the wall must be included to determine the "effective" eccentricity of the resultant load. Eccentricity is calculated by dividing the total applied moment by the total applied axial load.

Table 10-1 contains the formulas used for calculating the design axial load strength using the empirical design method. Note that walls must contain both horizontal and vertical reinforcement. The minimum area and the maximum spacing of reinforcement are prescribed in ACI 14.3. The length of wall to be considered effective for each concentrated load (beam reaction) must not exceed the center-to-center distance between loads, nor the width of bearing plus 4 times the thickness (ACI 14.2.4). The wall must be anchored to the footing, floors, and roof (ACI 14.2.6).

Walls designed as compression members

Where wall geometry and loading conditions do not satisfy the limitations of ACI 14.5 (usually where lateral loads are present or floors are supported on ledgers, and for grid walls), walls are designed as compression members, as described in ACI 14.4. ACI 14.4 refers the designer to ACI Chapter 10 (10.2, 10.3, 10.10, 10.11, 10.12, 10.13, 10.14, 10.17). The provisions of ACI 14.2 and 14.3 also apply. In addition, ties on vertical wall reinforcement are not required (as for columns) as long as the vertical reinforcement area is not greater than 0.01 times the gross concrete area (ACI 14.3.6).

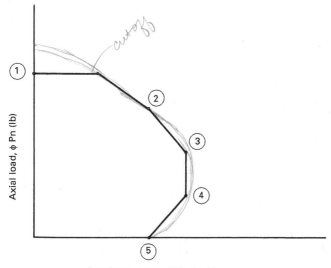

Figure 10-2 Transition stages on interaction diagram.

In general, walls must be designed for the combined effects of axial load and bending moment. This is usually accomplished by plotting an axial load-bending moment interaction diagram, as shown in Figure 10-2. Design aids like this facilitate the selection of reinforcement.

For simplicity, each design curve is plotted with straight lines connecting a number of points corresponding to certain transition stages. In general, the transition stages are defined as follows (refer to Figure 10-2):

Stage 1 Pure compression (no bending moment)

Stage 2 Stress in reinforcement closest to tension face = 0.

Stage 3 Stress in reinforcement closest to tension face = 0.5 times the yield stress.

Stage 4 Balanced point; stress in reinforcement closest to tension face = yield stress.

Stage 5 Pure bending (no axial load).

A designer can construct such an axial load-bending moment interaction diagram by calculating the nominal axial load strength and the corresponding nominal flexural strength for each of the stages and connecting the points with straight lines. The designer would then calculate the factored axial load and moment, plot this point on the interaction diagram, and if it falls within the region bound by the interaction diagram, the design is adequate.

Fortunately, this process is greatly simplified for most residential walls since the factored axial loads are typically below the balanced point (stage 4). Therefore the designer only needs to construct the interaction diagram between stages 4 and 5.

Table 10-2 lists the formulas to calculate stage 4 (balanced point). Figure 10-3 illustrates the balanced strain conditions. The extreme fiber in compression reaches a strain of 0.003 at the same time the steel reaches the yield strain. Table 10-3 gives the formula required for calculating stage 5 (no axial load).

Once these two points are calculated for a particular wall cross section and area of steel, they can be plotted on a graph, as shown in Figure 10-4. The two points can be connected with a straight line, and the graph becomes the interaction diagram for that wall cross section with that amount of reinforcement. The designer could develop interaction curves for several wall geometries and areas of steel and use them as a design aid.

It should be noted that the interaction curve developed using only two points connected with a straight line is only approximate. The actual interaction curve can be developed by plotting several points between stages 4 and 5. However, the straight-line approximation is adequate for purposes of design.

Slenderness

The design of many walls must also consider slenderness. Two methods for slenderness considerations are specified in ACI 318-95. A second-order analysis, which takes into account variable wall stiffness as well as the effects of member curvature and lateral drift, duration of the loads, shrinkage and creep, and interaction with the supporting foundation, is specified in ACI 10.10.1. In lieu of that procedure, the approximate evaluation of slenderness effects prescribed in ACI 10.11 may be used.

The approximate method uses the moment magnifier concept to account for slenderness effects. Moments are computed using ordinary first-order analysis and multiplied by a "moment magnifier," which is a function of the factored axial load P_u and the critical buckling load P_c for the axial member. Nonsway and sway frames are treated separately in ACI 10.12 and 10.13. Provisions applicable to both nonsway and sway frames are given in ACI 10.11.

The moment magnifier method requires the designer to distinguish between nonsway and sway frames. Frequently this can be done by inspection, comparing the total lateral stiffness of the compression

TABLE 10-2 Calculations for Stage 4 of Interaction Diagram

$$c = \left(\frac{\epsilon_c}{\epsilon_c + \epsilon_y}\right)d \quad \text{(ACI 10.2.2)}$$

$$\epsilon_c = 0.003 \quad \text{(ACI 10.2.3)}$$

$$\epsilon_y = \frac{f_y}{E_s} \quad \text{(ACI 10.2.4)}$$

$$a = \beta_1 c \quad \text{(ACI 10.2.7.1)}$$

$$C_c = 0.85 a b f_c' \quad \text{(ACI 10.2.7.1)}$$

$$T_s = A_s f_y \quad \text{(ACI 10.2.4)}$$

$$\phi P_b = \phi(C_c - T_s)$$

$$\phi M_b = \phi C_c \left(d - \frac{a}{2}\right)$$

Variable	Identity	Unit
a	Depth of equivalent rectangular stress block*	in
A_s	Area of tension reinforcement	in^2
b	Width of compression face	in
c	Distance from extreme compression fiber to neutral axis	in
C_c	Compression in concrete	lb
d	Distance from extreme compression fiber to centroid of tension reinforcement	in
E_s	Modulus of elasticity of reinforcement†	psi
f_c'	Specified compressive strength of concrete	psi
f_y	Specified yield strength of reinforcement	psi
M_b	Nominal flexural strength at balanced strain conditions‡	in · lb
P_b	Nominal axial load strength at balanced strain conditions‡	lb
T_s	Tension in reinforcement	lb
β_1	Factor¶	dimensionless
ϵ_c	Strain in concrete	dimensionless
ϵ_y	Yield strain of reinforcing steel	dimensionless
ϕ	Strength reduction factor§	dimensionless

*See ACI 10.2.7.1.
†Set equal to 29,000,000 psi (ACI 8.5.2).
‡See ACI 10.3.2.
¶Defined in ACI 10.2.7.3; equal to 0.85 for f_c' up to and including 4000 psi.
§Set equal to 0.70 for combined axial load and flexure (ACI 9.3.2).

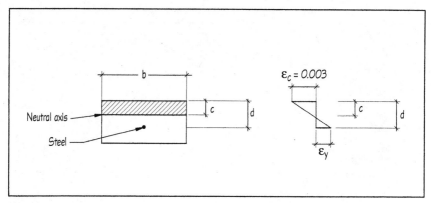

Figure 10-3 Balanced strain conditions.

TABLE 10-3 Calculations for Stage 5 of Interaction Diagram (No Axial Load)

$$a = \frac{A_s f_y}{0.85 f_c' b}$$

$$\phi M_n = \phi A_s f_y \left(d - \frac{a}{2} \right)$$

Variable	Identity	Unit
a	Depth of equivalent rectangular stress block*	in
A_s	Area of tension reinforcement	in²
b	Width of compression face	in
d	Distance from extreme compression fiber to centroid of tension reinforcement	in
f_c'	Specified compressive strength of concrete	psi
f_y	Specified yield strength of reinforcement	psi
M_n	Nominal flexural strength at condition of no axial load	in · lb
ϕ	Strength reduction factor†	dimensionless

*See ACI 10.2.7.1.
†Set equal to 0.90 for flexure (ACI 9.3.2).

member to that of the bracing elements. A compression member may be assumed braced if it is located in a story in which bracing elements (shear walls or other types of lateral bracing) have such substantial stiffness to resist the lateral deflections of the story that any resulting lateral deflection is not large enough to affect the column strength substantially. Most homes built with flat or grid ICF systems on all four sides could be categorized as nonsway frames.

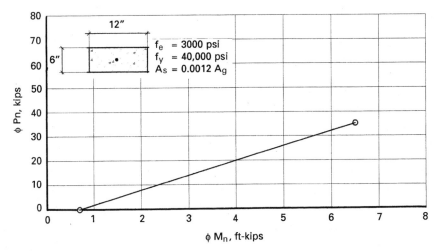

Figure 10-4 Load moment interaction diagram for 6-inch flat wall with $A_s = 0.0012 A_g$.

Table 10-4 lists the equations required for calculating the magnified moments for nonsway frames. Before calculating magnified moments the designer must check that the wall is slender. ACI 10.12.2 stipulates that slenderness can be ignored if

$$\frac{kl_u}{r} < 34 - 12 \frac{M_1}{M_2} \qquad \text{(ACI EQ 10-8)}$$

where M_1 = smaller factored end moment
M_2 = larger factored end moment
k = effective length factor, = 1.0 (ACI 10.12.1)
l_u = unsupported length of compression member
r = radius of gyration of cross section, = 0.3 times the thickness of rectangular members (ACI 10.11.2)

and M_1/M_2 is positive if the wall is bent in single curvature. M_1/M_2 cannot be taken less than -0.5. Very often, M_1 is assumed to be 0, and ACI EQ 10-8 reduces to

$$\frac{kl_u}{r} < 34$$

It is important to note that ACI EQs 10-12 and 10-13 for the flexural stiffness EI of compression members were not derived for members with a single layer of reinforcement. For members with a single layer

TABLE 10-4 Formulas for Magnified Moments in Nonsway Frames

$$M_c = \delta_{ns} M_2 \qquad \text{(ACI EQ 10-9)}$$

$$\delta_{ns} = \frac{C_m}{1 - P_u/0.75 P_c} \geq 1.0 \qquad \text{(ACI EQ 10-10)}$$

$$P_c = \frac{\pi^2 EI}{(kl_u)^2} \qquad \text{(ACI EQ 10-11)}$$

$$C_m = 0.6 + 0.4 \frac{M_1}{M_2} \geq 0.4 \qquad \text{(ACI EQ 10-14)}$$

or

$C_m = 1.0$ for members with transverse loads between supports

$$M_{2,\min} = P_u(0.6 + 0.03h) \qquad \text{(ACI EQ 10-15)}$$

Variable	Identity	Unit
C_m	Factor relating moment diagram to equivalent uniform moment diagram	dimensionless
E_c	Modulus of elasticity of concrete	psi
EI	Flexural stiffness of compression member*	lb · in²
I_g	Moment of inertia of gross concrete section	in⁴
k	Effective length factor	dimensionless
l_u	Unsupported length of compression member	in
M_1/M_2	Ratio of smaller factored end moment to larger factored end moment; positive if compression member is bent in single curvature	dimensionless
M_2	Larger factored end moment, always positive	in · lb
$M_{2,\min}$	Minimum value of M_2	in · lb
M_c	Magnified factored moment to be used for designing compression members	in · lb
P_c	Critical load	lb
P_u	Factored axial load	lb
δ_{ns}	Moment magnification factor for nonsway frames	dimensionless

*See Table 10-5.

of reinforcement, notes on ACI 318-95, published by the Portland Cement Association (Ghosh, Fanella, and Rabbat, 1996), suggest using the formulas in Table 10-5 to calculate EI.

Perpendicular shear in walls

The design for shear forces perpendicular to the wall is governed by ACI 11.10 for reinforced concrete. This clause refers the designer to ACI 11.12, which in turn refers to ACI 11.1 through 11.5. Although ACI 11.1 through 11.5 permit the use of shear reinforcement (stir-

TABLE 10-5 Formulas for Flexural Stiffness of Compression Members with One Layer of Reinforcement

$$EI = \frac{E_c I_g}{\beta}\left(0.5 - \frac{e}{h}\right) \geq 0.1\frac{E_c I_g}{\beta}$$

$$\leq 0.4\frac{E_c I_g}{\beta}$$

$$\beta = 0.9 + 0.5\beta_d^2 - 12\rho \geq 1.0$$

Variable	Identity	Unit
e	Eccentricity of axial load	in
EI	Flexural stiffness of compression member	lb · in²
E_c	Modulus of elasticity of concrete	psi
h	Thickness of wall	in
I_g	Moment of inertia of gross concrete section	in⁴
β	As defined in equation	dimensionless
β_d	Ratio of dead load to total load	dimensionless
ρ	Ratio of area of vertical reinforcement to gross concrete area	dimensionless

rups) for walls, it is impractical to use stirrups for thin walls and should be avoided. The designer is advised to increase the thickness of the wall or the specified compressive strength to increase shear strength, if required. Fortunately, shear forces are typically small and will rarely govern wall thickness or concrete compressive strength.

Table 10-6 presents the formulas and variables for calculating the nominal shear strength of a wall. Formulas for calculating the shear strength provided by reinforcement are not included here, but can be found in ACI 11.5.

If reinforcement is required, the designer should consider using the shear-friction method found in ACI 11.7, in which vertical reinforcement can be used to resist shear forces. This method is particularly useful for calculating shear strength at a plane of weakness, such as a construction joint or cold joint.

There is an important variation in shear strength calculations across different classes of ICF systems. It is the assumption used to determine the value of web width b_w. Figure 10-5 depicts the alternatives. With a flat wall the problem is straightforward since the wall section is constant. The web width is the same as the distance between reinforcing bars. Thus a wall having vertical reinforcement every 18 inches on center has a cross-sectional area (in square inches) of 18 times the concrete thickness. And since one of these 18-inch-wide reinforced vertical "members" occurs every 1.5 feet, one would calculate the shear capacity per lineal foot as ⅔ (i.e., ¹²⁄₁₈) times V_c.

TABLE 10-6 Calculation of Shear Strength Perpendicular to Wall for Reinforced Concrete

$\phi V_n \geq V_u$	(ACI EQ 11-1)
$V_n = V_c + V_s$	(ACI EQ 11-2)
$V_c = 2\sqrt{f_c'}\, b_w d$	(ACI EQ 11-3)

Variable	Identity	Unit
b_w	Web width, or diameter of circular section	in
d	Distance from extreme compression fiber to centroid of longitudinal tension reinforcement	in
f_c'	Specified compressive strength of concrete	psi
V_c	Nominal shear strength provided by concrete	lb
V_n	Nominal shear strength	lb
V_s	Nominal shear strength provided by shear reinforcement*	lb
V_u	Factored shear force at section	lb
ϕ	Strength reduction factor†	dimensionless

*To be set equal to 0 if no shear reinforcement is provided.
†To be set equal to 0.85 for shear (ACI 9.3.2).

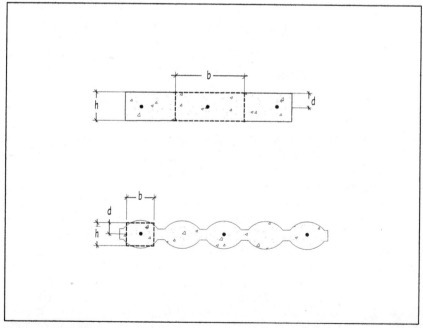

Figure 10-5 Assumed concrete cross-sectional dimensions for shear perpendicular to the wall for flat and grid walls.

More open to interpretation is how to calculate the web width in a grid wall. The conservative assumption, and the one most often used, is to calculate it across the reinforced vertical cavities only. The concrete in cavities between the reinforced cavities is assumed to make no structural contribution. This assumption may be conservative, and it limits the calculated shear capacity per lineal foot of grid walls.

Assumed dimensions are sometimes further simplified in grid systems. As also depicted in Figure 10-5, when the cross section of the post is an ellipse, for calculation purposes it is sometimes assumed to be a rectangle of equal or smaller dimensions than the actual ellipse. This simplifies several calculations (to be described later).

Parallel shear in walls

Design for shear parallel to the wall for reinforced concrete is governed by ACI 11.10. Shear parallel to the wall can be resisted by a combination of the shear strength provided by concrete and the shear strength provided by shear reinforcement, which consists of vertical and horizontal bars placed in the wall. The levels of shear are typically small enough to be resisted by the shear strength of the concrete alone.

Table 10-7 presents the formulas and variables for calculating the nominal shear strength of a wall. Formulas for the design of shear reinforcement for walls where the shear force V_u exceeds the shear strength ϕV_c are not included here, but can be found in ACI 11.10.9.

Figure 10-6 illustrates the appropriate dimension to use in calculating the nominal shear strength. Calculations are normally performed for a unit length of wall, usually 12 inches for flat walls or the center-to-center distance between cavities for grid walls. Dimensions are simplified for grid systems that have complex cross-sectional geometries. All posts, reinforced or not, are used in this calculation.

Since post-and-beam systems consist of true beams and columns, they are correctly designed as frame systems rather than shear wall systems (as described for flat and grid walls). Although not presented here, engineering procedures for frame systems are in most structural design texts.

Lintel bending

The design of lintels over openings for bending is governed by the provisions for flexural members (beams) outlined in ACI Chapter 10. The formulas used to calculate the nominal flexural strength of a lintel are listed in Table 10-8. Figure 10-7 illustrates the cross-sectional dimensional variables used for designing lintels in bending.

218   Engineering

TABLE 10-7   Calculation of Shear Strength Parallel to Wall for Reinforced Concrete

$\phi V_n \geq V_u$	(ACI 11.1)
$V_n = V_c + V_s$	(ACI 11.2)
$V_c = 2\sqrt{f_c'}\, hd$	(ACI 11.10.5)
$d = 0.8 l_w$	(ACI 11.10.4)

Variable	Identity	Unit
d	As in equation	in
f_c'	Specified compressive strength of concrete	psi
h	Overall thickness of wall	in
l_w	Length of wall	in
V_c	Shear strength provided by concrete	lb
V_n	Nominal shear strength	lb
V_s	Nominal shear strength provided by shear reinforcement*	lb
V_u	Factored shear force at section	lb
ϕ	Strength reduction factor†	dimensionless

*Assume equal to 0 when $\phi V_c \geq V_u$. If $V_u \geq \phi V_c$, design shear reinforcement using ACI 11.10.9.
†To be set equal to 0.85 for shear (ACI 9.3.2).

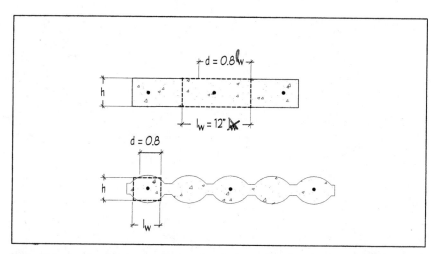

Figure 10-6   Assumed concrete cross-sectional dimensions for shear parallel to the wall for flat and grid walls.

Structural Design

TABLE 10-8 Calculation of Nominal Moment Strength of Lintels

$$\phi M_n \geq M_u$$

$$\phi M_n = \phi A_s f_y \left(d - \frac{a}{2} \right)$$

$$a = \frac{A_s f_y}{0.85 f_c' b}$$

Variable	Identity	Unit
a	Depth of equivalent rectangular stress block	in
A_s	Area of tension reinforcement	in²
b	Width of compression face of member	in
d	Distance from extreme compression fiber to centroid of tension reinforcement	in
f_c'	Specified compressive strength of concrete	psi
f_y	Specified yield strength of reinforcement	psi
M_n	Nominal moment strength	in · lb
M_u	Factored moment at section	in · lb
ϕ	Strength reduction factor*	dimensionless

*To be set equal to 0.9 for flexure (ACI 9.3.2).

Determining the appropriate width of the compression face b is straightforward for flat walls, but open to interpretation for grid and post-and-beam walls. For flat walls, b is equal to the width of the concrete wall. For grid walls, a rectangular approximation of the cross section can be used, as shown in Figure 10-7 (center). For some grid systems, especially those with foam ties penetrating the wall and post-and-beam systems, it might be necessary to use special lintel blocks or modified standard blocks to create solid cross sections for longer spans. The width of the compression face is determined by using a rectangular approximation of the cross section, as shown in Figure 10-7 (bottom).

Lintel shear

The design of lintels for shear is governed by the provisions for shear in ACI Chapter 11. The formulas used to calculate the nominal shear strength of a lintel are listed in Table 10-9. Figure 10-8 illustrates the cross-sectional dimensional variables used for designing lintels in shear.

Determining the appropriate web width b_w is straightforward for flat walls, but is open to interpretation for grid and post-and-beam

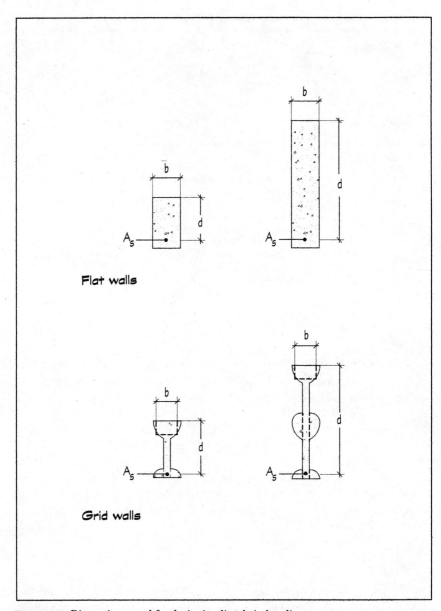

Figure 10-7 Dimensions used for designing lintels in bending.

TABLE 10-9 Calculation of Nominal Shear Strength of Lintels

$\phi V_n \geq V_u$	(ACI 11.1)
$V_n = V_c + V_s$	(ACI 11.2)
$V_c = 2\sqrt{f_c'}\, b_w d$	(ACI 11.3)
$V_s = \dfrac{A_v f_y d}{s}$	(ACI EQ 11-5)
$A_{v,\min} = \dfrac{50 b_w s}{f_y}$ when $V_u > 0.5\phi V_c$	(ACI EQ 11-13, 11.5.5.3)
$s \leq d/2 \leq 24$	(ACI 11.5.4.1)
$s \leq d/4 \leq 12$ when $V_s > 4\sqrt{f_c'}\, b_w d$	(ACI 11.5.4.3)

Variable	Identity	Unit
A_v	Area of shear reinforcement within distance s	in²
$A_{v,\min}$	Minimum area of shear reinforcement	in²
b_w	Web width	in
d	Distance from extreme compression fiber to centroid of longitudinal tension reinforcement	in
f_c'	Specified compressive strength of concrete	psi
f_y	Specified yield strength of reinforcement	psi
s	Spacing of shear reinforcement	in
V_c	Nominal shear strength provided by concrete	lb
V_n	Nominal shear strength	lb
V_s	Nominal shear strength provided by shear reinforcement*	lb
V_u	Factored shear force	lb
ϕ	Strength reduction factor†	dimensionless

*To be set equal to 0 if no reinforcement is provided.
†To be set equal to 0.85 for shear (ACI 9.3.2).

walls. For flat walls, b_w is equal to the wall width. For grid walls, however, it should be taken as the smallest thickness of concrete within the web. Therefore, for grid walls with foam ties and for post-and-beam walls there is no effective way of producing a deep lintel for long spans without using special lintel forms or by modifying regular forms to create a solid lintel. Once a solid lintel is created, the web width can be taken as the least dimension of the web thickness (Figure 10-8). It is also sometimes possible to employ conventional formwork (as discussed in Chapter 9) to increase the web width.

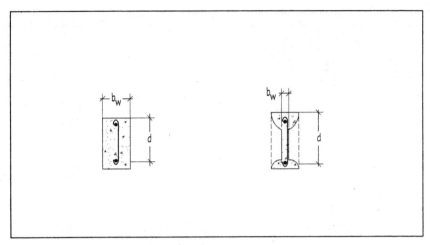

Figure 10-8 Dimensions used for designing lintels in shear.

Reinforced Concrete Example

Figure 10-9 presents dimensions of a hypothetical two-story house on which we perform calculations for ICF grid exterior walls. It also includes dimensions of the assumed grid. As noted in the foregoing, calculations for flat walls would be similar, although simpler in certain respects.

Table 10-10 contains assumed loads, properties of the materials used, and reinforcement size and spacing. The reinforcement variables are starting assumptions, which may be adjusted up or down as the calculations indicate.

Shear perpendicular to wall

The calculations in Table 10-11 check the capacity of the wall, as designed, to resist the shear force perpendicular to the wall. The numbers used pertain to the lower story, where the force will be greatest. As noted, the results show that the wall strength far exceeds requirements.

Shear parallel to wall

Figure 10-10 depicts the shear forces parallel to the wall. We focus on the south wall because, by inspection, it will experience the highest shear per lineal foot.

Table 10-12 contains the check of the wall's adequacy to resist the parallel shear. Again numbers pertain to the lower story, where the total force will be greatest. Note that 22 feet of wall (total length minus the 4 feet containing openings) is available to resist shear.

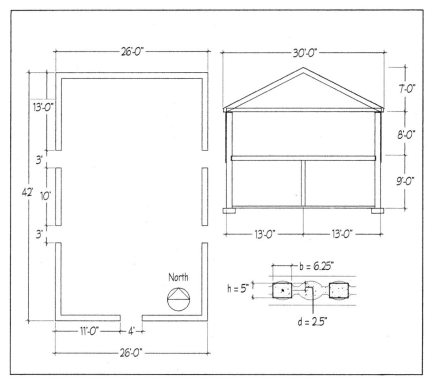

Figure 10-9 Hypothetical house and assumed wall cross section for reinforced concrete example.

TABLE 10-10 Load and Materials Assumptions for Reinforced Concrete Example

Variable	Assumed value
Wind load	±25 psf
Roof live load	20 psf
Roof dead load	20 psf
Floor live load	40 psf
Floor dead load	10 psf
Vertical reinforcement size and spacing	One #4 @ 24" oc
f_c' (specified concrete compressive strength)	3000 psi
f_y (specified reinforcement tensile strength)	40,000 psi

TABLE 10-11 Check of Shear Force Perpendicular to Wall for Reinforced Concrete Example

Lower story:

$$V = (25 \text{ psf})(2 \text{ ft})\left(\frac{9 \text{ ft}}{2}\right) = 225 \text{ lb}$$

Factored shear force:

$U_3 = 0.9D + 1.3W$ (ACI 9.2.2)

$V_u = 1.3 \times 225 = 293 \text{ lb}$

$\phi V_c = 0.85(2)\sqrt{3000}(6.25)(2.5) = 1455 \text{ lb}$

$\phi V_c \geq V_u$ OK

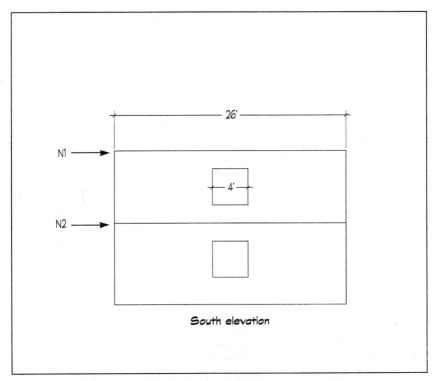

Figure 10-10 Shear forces parallel to the south wall in reinforced concrete example.

TABLE 10-12 Check of Shear Force Parallel to South Wall for Reinforced Concrete Example

$$N_1 = (25 \text{ psf})\left(7 \text{ ft} + \frac{8 \text{ ft}}{2}\right)\left(\frac{42 \text{ ft}}{2}\right) = 5775 \text{ lb*}$$

$$N_2 = (25 \text{ psf})\left(\frac{8 \text{ ft}}{2} + \frac{9 \text{ ft}}{2}\right)\left(\frac{42 \text{ ft}}{2}\right) = 4463 \text{ lb*}$$

Shear in lower story:

$$N_1 + N_2 = 5775 + 4463 = 10{,}238 \text{ lb}$$

$$V_u = 1.3 \times 10{,}238 \text{ lb} = 13{,}309 \text{ lb}$$

$$= \frac{13{,}309}{22} = 605 \text{ lb/post}$$

$$\phi V_c = 0.85(2)\sqrt{3000}(5)(0.8)(6.25) = 2328 \text{ lb/post}$$

$$\phi V_c \geq V_u \quad \text{OK}$$

*N_1 is the shear force applied to the top story parallel to the south wall, and N_2 the shear force applied to the bottom story parallel to the south wall.

With the assumed grid system (having one vertical post per lineal foot) this corresponds to 22 individual vertical concrete members resisting shear.

Axial and flexural loading

Figure 10-11 depicts dimensions important to calculating the interaction diagrams for combined axial and flexural loads. As drawn, the vertical posts of the grid wall are assumed to have a rectangular cross section. Tables 10-13 and 10-14 contain the preliminary calculations necessary to produce interaction diagrams. They cover the balanced condition and the pure moment (no axial load) condition, respectively.

Table 10-15 contains calculations of the bending moments and the axial loads in the bottom story. Table 10-16 presents a check of the bottom story wall for slenderness and finds that it must be considered. Tables 10-17 through 10-19 present calculations of factored axial load and bending moment for the bottom story.

The resulting completed diagram is shown in Figure 10-12. The three points U_1, U_2, and U_3, representing the three load cases, fall within the interaction diagram. Therefore the design is adequate. By inspection one can determine that the top story will also be adequate.

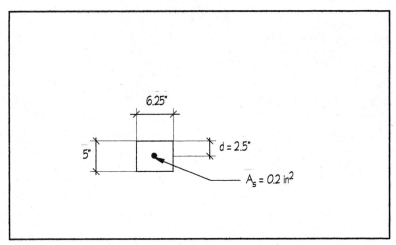

Figure 10-11 Dimensions of concrete post for calculation of interaction diagram.

TABLE 10-13 Bottom-Story Balanced Condition for Interaction Diagram in Reinforced Concrete Example

$c = \left(\dfrac{\epsilon_c}{\epsilon_c + \epsilon_y} \right) d$

$\epsilon_c = 0.003$

$\epsilon_y = \dfrac{40{,}000}{29{,}000{,}000} = 0.0014$

$c = \left(\dfrac{0.003}{0.003 + 0.0014} \right)(2.5) = 1.7 \text{ in}$

$a = 0.85(1.7) = 1.45 \text{ in}$

$C_c = 0.85\, abf_c' = 0.85(1.45)(6.25)(3000) = 23{,}109 \text{ lb}$

$T_s = A_s f_y = 0.2(40{,}000) = 8000 \text{ lb}$

$\phi P_b = \phi(C_c - T_s) = 0.7(23{,}109 - 8000) = 10{,}576 \text{ lb}$

$\phi M_b = \phi C_c \left(d - \dfrac{a}{2} \right) = 0.7\left[23{,}109 \left(2.5 - \dfrac{1.45}{2} \right) \right] = 28{,}713 \text{ in} \cdot \text{lb}$

Structural Design 227

TABLE 10-14 Bottom-Story Pure Moment (No Axial Load) Condition for Reinforced Concrete Example

$$a = \frac{A_s f_y}{0.85 f_c' b} = \frac{0.2(40,000)}{0.85(6.25)(3000)} = 0.502 \text{ in}$$

$$\phi M_n = \phi A_s f_y \left(d - \frac{a}{2}\right) = 0.9(0.2)(40,000)\left(2.5 - \frac{0.502}{2}\right) = 16,193 \text{ in} \cdot \text{lb}$$

TABLE 10-15 Calculation of Bending Moments and Axial Loads for Bottom Story in Reinforced Concrete Example

Dead loads:

$$P_{\text{roof}} = (15 \text{ ft})(20 \text{ psf}) = 300 \text{ plf}$$

$$P_{\text{floor}} = \left(\frac{13 \text{ ft}}{2}\right)(10 \text{ psf}) = 65 \text{ plf}$$

$$P_{\text{wall}} = \left(8 + \frac{9}{2}\right)(56) = 700 \text{ plf @ midheight}$$

$$P_{\text{wall}} = (8 + 9)(56) = 952 \text{ plf @ bottom of wall}$$

$$M_{\text{floor}} = \frac{(65 \text{ plf})(3 \text{ in})}{2} = 97.5 \text{ in} \cdot \text{lb @ midheight (assuming eccentricity of 3")}$$

Live loads:

$$P_{\text{roof}} = (15 \text{ ft})(20 \text{ psf}) = 300 \text{ plf}$$

$$P_{\text{floor}} = \left(\frac{13 \text{ ft}}{2}\right)(40 \text{ psf}) = 260 \text{ plf}$$

$$M_{\text{floor}} = \frac{260(3)}{2} = 390 \text{ in} \cdot \text{lb @ midheight}$$

Wind loads:

$$M_{\text{wind}} = \frac{25(9)^2}{8} = 253 \text{ ft} \cdot \text{lb} = 3038 \text{ in} \cdot \text{lb}$$

TABLE 10-16 Check for Slenderness in Bottom Story of Reinforced Concrete Example

$$\frac{k l_u}{r} = \frac{1(9 \times 12)}{0.3(5)} = 72 > 34 \quad \text{Hence, slenderness must be considered.}$$

TABLE 10-17 Load Case 1 Calculations for Bottom Story in Reinforced Concrete Example

$U_1 = 1.4D + 1.7L$

$P_u = [1.4(300 + 65 + 952) + 1.7(300 + 260)]2$

$\quad = 5592$ lb @ bottom of wall

$M_u = 0$ @ bottom of wall

or

$M_u = M_{2,\min} = P_u[0.6 + 0.03(5)] = 5592(0.75) = 4194$ in · lb

$M_2 = 4194$ in · lb

$E_c = 3{,}122{,}019$ psi

$I_g = 65.1$ in^4

$e = \dfrac{M}{P} = \dfrac{4194}{5592} = 0.75$

$h = 5$ in

$\beta_d = \dfrac{3688}{5592} = 0.66$

$\rho = 0.0064$

$\beta = 0.9 + 0.5(0.66)^2 - 12(0.0064) = 1.04$

$EI = \left(0.5 - \dfrac{0.75}{5}\right)\dfrac{E_c I_g}{\beta} = 0.35\dfrac{E_c I_g}{\beta}$

$EI = 0.35\left[\dfrac{3{,}122{,}019(65.1)}{1.04}\right] = 68{,}399{,}234$ lb · in^2

$P_c = \dfrac{\pi^2 EI}{(kl_u)^2} = \dfrac{\pi^2(68{,}399{,}234)}{[1(9 \times 12)]^2} = 57{,}877$ lb

$C_m = 0.6 + 0.4\dfrac{M_1}{M_2} = 0.6$

$\delta_{ns} = \dfrac{0.6}{5592 / [0.75(57{,}877)]} = 0.69 \geq 1.0$

$\delta_{ns} = 1.0$

$M_c = \delta_{ns} M_2 = 1.0(4194) = 4194$ in · lb

Hence, plot $\phi P_u = 5592$ lb and $\phi M_c = 4194$ in · lb on the interaction diagram. OK.

TABLE 10-18 Load Case 2 Calculations for Bottom Story in Reinforced Concrete Example

$U_2 = 0.75(1.4D + 1.7L + 1.7W) = 1.05D + 1.275L + 1.275W$

$P_u = [1.05(300 + 65 + 700) + 1.275(300 + 260)](2 \text{ ft}) = 3665 \text{ lb}$

$M_u = [1.05(97.5) + 1.275(390) + 1.275(3038)](2 \text{ ft}) = 8946 \text{ in} \cdot \text{lb}$

or

$M_u = M_{2,\min} = 3665(0.75) = 2749 \text{ in} \cdot \text{lb}$

$M_2 = 8946 \text{ in} \cdot \text{lb}$

$E_c = 3{,}122{,}019 \text{ psi}$

$I_g = 65.1 \text{ in}^4$

$e = \dfrac{8946}{3665} = 2.44 \text{ in}$

$h = 5 \text{ in}$

$\beta_d = \dfrac{2237}{3665} = 0.610$

$\rho = 0.0064$

$\beta = 0.9 + 0.5(0.610) - 12(0.0064) = 1.13$

$EI = \left(0.5 - \dfrac{2.44}{5}\right)\dfrac{E_c I_g}{\beta} = 0.012 \dfrac{E_c I_g}{\beta} = 0.1 \dfrac{E_c I_g}{\beta}$

$= 0.1\left[\dfrac{3{,}122{,}019(65.1)}{1.13}\right] = 17{,}986{,}145 \text{ lb} \cdot \text{in}^2$

$P_c = \dfrac{\pi^2 EI}{(kl_u)^2} = \dfrac{\pi^2(17{,}986{,}145)}{[1(9 \times 12)]^2} = 15{,}219 \text{ lb}$

$C_m = 1.0$

$\delta_{ns} = \dfrac{1.0}{1 - 3665/[0.75(15{,}219)]} = 1.47$

$M_c = 1.47(8946) = 13{,}151 \text{ in} \cdot \text{lb}$

Hence, plot $\phi P_u = 3665$ lb and $\phi M_c = 13{,}151$ in · lb on the interaction diagram. OK.

TABLE 10-19 Load Case 3 Calculations for Bottom Story in Reinforced Concrete Example

---

$U_3 = 0.9D + 1.3W$

$P_u = [0.9(300 + 65 + 700)](2 \text{ ft}) = 1917 \text{ lb}$

$M_u = [0.9(97.5) + 1.3(3038)](2 \text{ ft}) = 8074 \text{ in} \cdot \text{lb}$

or

$M_u = M_{2,\text{min}} = 1917(0.75) = 1438 \text{ in} \cdot \text{lb}$

$M_2 = 8074 \text{ in} \cdot \text{lb}$

$E_c = 3{,}122{,}019 \text{ psi}$

$I_g = 65.1 \text{ in}^4$

$e = \dfrac{8074}{1917} = 4.21 \text{ in}$

$h = 5 \text{ in}$

$\beta_d = 1.0$

$\rho = 0.0064$

$\beta = 0.9 + 0.5(1)^2 - 12(0.0064) = 1.32$

$EI = \left(0.5 - \dfrac{4.12}{5}\right)\dfrac{E_c I_g}{\beta} = -0.32\dfrac{E_c I_g}{\beta} = 0.1\dfrac{E_c I_g}{\beta}$

$= 0.1\left[\dfrac{3{,}122{,}019(65.1)}{1.32}\right] = 15{,}397{,}230 \text{ lb} \cdot \text{in}^2$

$P_c = \dfrac{\pi^2 EI}{(kl_u)^2} = \dfrac{\pi^2(15{,}397{,}230)}{[1(9 \times 12)]^2} = 13{,}029 \text{ lb}$

$C_m = 1.0$

$\delta_{ns} = \dfrac{1.0}{1 - 1917/[0.75(13{,}029)]} = 1.24$

$M_c = 1.24(8074) = 10{,}012 \text{ in} \cdot \text{lb}$

Hence, plot $\phi P_u = 1917 \text{ lb}$ and $\phi M_c = 10{,}012 \text{ in} \cdot \text{lb}$ on the interaction diagram.   OK.

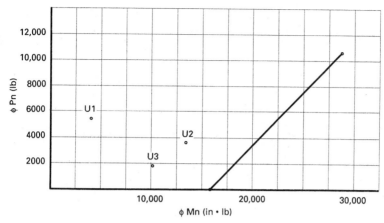

Figure 10-12 Completed interaction diagram.

Structural Plain Concrete Design Procedures

By definition, structural plain concrete is concrete that either is unreinforced or contains less reinforcement than the minimum amount specified for reinforced concrete. Since the structural integrity of structural plain concrete members depends solely on the properties of the concrete, ACI 318 limits the use of structural plain concrete to members that are continuously supported by soil or by other structural members capable of providing vertical support continuously throughout the length of the plain concrete member; members in which arch action assures compression under all conditions of loading; and walls and pedestals. ACI Chapter 22 contains specific design provisions for structural plain concrete walls, footings, and pedestals.

ACI 22.3.1 stipulates that structural plain concrete members must be divided into flexurally discontinuous elements with contraction or isolation joints. The size of each element shall be limited to control the buildup of excessive internal stresses within each element caused by restraint to movements from creep, shrinkage, and temperature effects. The commentary to ACI 318 suggests that joints may not be required where random cracking due to creep, shrinkage, and temperature effects will not affect the structural integrity. It is probably reasonable to assume that ICF walls fall into this category and therefore could be constructed without contraction joints and designed as structural plain concrete.

ACI Chapter 22 does not have a minimum requirement for vertical and horizontal reinforcement. Like ACI Chapter 14, however, it does have minimum requirements for reinforcement around windows.

TABLE 10-20 Formulas for Calculation of Shear Strength of Structural Plain Concrete Walls

$\phi V_n \geq V_u$	(ACI EQ 22-7)
$V_n = \tfrac{4}{3}\sqrt{f_c'}\, bh$	(ACI EQ 22-8)

Variable	Identity	Unit
b	Width of member	in
h	Thickness of member	in
f_c'	Specified compressive strength of concrete	psi
V_n	Nominal shear strength at section	lb
V_u	Factored shear force at section	lb
ϕ	Strength reduction factor*	dimensionless

*To be set equal to 0.65 for structural plain concrete (ACI 9.3.5).

Perpendicular and parallel shear in walls

The design for shear in walls is governed by ACI 22.5.4 for structural plain concrete. Perpendicular and parallel shear is resisted by shear strength provided by the concrete alone. Table 10-20 presents the formulas used to calculate the nominal shear capacity. Figure 10-13 illustrates the appropriate dimensions to use in this calculation.

Axial and flexural loading

ACI Chapter 22 offers two alternatives for designing plain concrete walls. The empirical design method is used for walls of solid rectangular cross sections, where the resultant of all factored axial loads falls within the middle one-third of the overall thickness of the wall. In addition, the thickness of the wall shall not be less than 1/24 the unsupported height or length, whichever is shorter, nor less than 5½ inches. Basement walls cannot be less than 7½ inches when designed using the empirical design method. For walls meeting these criteria, the designer can use the formula for nominal axial load strength P_{nw} in Table 10-21.

The second method, which can be used under all loading conditions, is the strength design method. It is required when the cross section is not solid and rectangular, or the eccentricity of the axial load is outside the middle third of the wall, or both. The same limits for thickness apply to the strength design method. Under this method, two interaction equations must be satisfied. They are given in Table 10-22. Note that all walls must be designed for a minimum eccentricity of $0.10h$ (ACI 22.6.3).

Walls designed by either method must be braced against lateral translation at the supports. In addition, not less than two #5 reinforc-

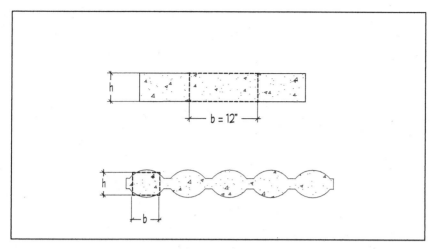

Perpendicular shear

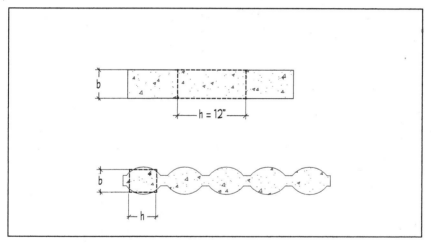

Parallel shear

Figure 10-13 Assumed concrete cross-sectional dimensions for shear in structural plain concrete walls.

ing bars shall be provided around all window and door openings, and they must extend 24 inches beyond the corners of the openings.

Applicability

The structural plain concrete provisions of ACI 318 are ideal for ICF walls in low-rise construction. The geometry of the walls and the

TABLE 10-21 Formulas for Nominal Axial Load Strength of Structural Plain Concrete Wall Using the Empirical Design Method

$$\phi P_{nw} \geq P_u \qquad \text{(ACI EQ 22-12)}$$

$$P_{nw} = 0.45 f_c' A_g \left[1 - \left(\frac{l_c}{32h} \right)^2 \right] \qquad \text{(ACI EQ 22-13)}$$

Variable	Identity	Unit
A_g	Gross area of section	in^2
f_c'	Specified compressive strength of concrete	psi
h	Thickness of member	in
l_c	Vertical distance between supports	in
P_{nw}	Nominal axial load strength	lb
P_u	Factored axial load	lb
ϕ	Strength reduction factor*	dimensionless

*Set equal to 0.65 for structural plain concrete (ACI 9.3.5).

TABLE 10-22 Interaction Equations for Structural Plain Concrete Walls Subjected to Flexure and Axial Load

$$\frac{P_u}{\phi P_n} + \frac{M_u}{\phi M_n} \leq 1 \qquad \text{(ACI EQ 22-5)}$$

where

$$M_n = 0.85 f_c' S$$

$$P_n = 0.6 f_c' \left[1 - \left(\frac{l_c}{32h} \right)^2 \right] A_1 \qquad \text{(ACI EQ 22-4)}$$

$$\frac{M_u}{S} - \frac{P_u}{A_g} \leq 5 \phi \sqrt{f_c'} \qquad \text{(ACI EQ 22-6)}$$

$$M_{u,\min} = 0.1 h P_u \qquad \text{(ACI 22.6.3)}$$

Variable	Identity	Unit
A_1	Loaded area	in^2
A_g	Gross area of section	in^2
f_c'	Specified compressive strength of concrete	psi
h	Overall thickness of member	in
l_c	Vertical distance between supports	in
M_n	Nominal moment strength at section	in · lb
M_u	Factored moment at section	in · lb
$M_{u,\min}$	Minimum factored moment*	in · lb
S	Elastic section modulus of section†	cu in
ϕ	Strength reduction factor‡	dimensionless

*Wall must be designed for an eccentricity corresponding to the maximum moment that can accompany the axial load but not less than $0.10h$ (ACI 22.6.3).
†$S = bh^2/6$ for rectangular sections.
‡To be set equal to 0.65 for structural plain concrete (ACI 9.3.5).

loading conditions are such that the walls can be designed to resist the loads based on the strength of the concrete alone. The designer could design the walls as structural plain concrete members and provide the required reinforcement at the openings.

Lintels must be designed as reinforced concrete members with the appropriate flexural and shear reinforcement. Portions of the wall on either side of an opening should also be checked.

It should be noted that the three model building codes (BOCA, SBCCI, and ICBO) place limits on the use of structural plain concrete design in areas subject to high seismic risk. BOCA and SBCCI allow its use for seismic performance categories A and B with no restrictions, and for category C with some minimum reinforcement requirements. Structural plain concrete design is not permitted in seismic performance categories D and E. The ICBO *Uniform Building Code* allows the use of structural plain concrete design in seismic zones 0 and 1, but not in zones 2, 3, or 4.

Structural Plain Concrete Example

Figure 10-14 contains dimensions of a hypothetical two-story house and flat wall that we assume in conducting engineering calculations according to the structural plain concrete procedure. In Table 10-23 loads, materials properties, and reinforcement parameters are assumed.

The calculations of Table 10-24 yield the various loads on the structure. Tables 10-25 through 10-27 check the capacity of the bottom story wall to withstand the applied loads. The lower story is selected for checking because it will be subject to the greater loadings.

As noted in the tables, under all calculations the lower walls are adequate to resist the loads as structural plain concrete walls.

236 Engineering

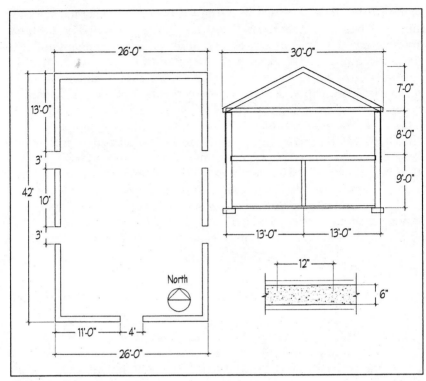

Figure 10-14 Hypothetical house and assumed wall cross section for structural plain concrete example.

TABLE 10-23 Load, Materials, and Dimension Assumptions for Structural Plain Concrete Example

Variable	Assumed value
Wind load	±25 psf
Roof live load	20 psf
Roof dead load	20 psf
Floor live load	40 psf
Floor dead load	10 psf
Weight of wall, $= 6/12(150 \text{ pcf})$	75 psf
f_c' (specified concrete compressive strength)	4000 psi
$S = \dfrac{bh^2}{6} = \dfrac{12(6)^2}{6}$	72 in^2
$A = bh = 12(6)$	72 in^2

TABLE 10-24 Bending Moments and Axial Loads for Bottom Story in Structural Plain Concrete Example

Dead loads:

$$P_{roof} = (15 \text{ ft})(20 \text{ psf}) = 300 \text{ plf}$$

$$P_{floor} = \frac{13 \text{ ft}}{2}(10 \text{ psf}) = 65 \text{ plf}$$

$$P_{wall} = \left(8 + \frac{9}{2}\right)(75 \text{ psf}) = 938 \text{ plf @ midheight}$$

$$P_{wall} = (8 + 9)(75 \text{ psf}) = 1275 \text{ plf @ bottom of wall}$$

$$M_{floor} = \frac{(65 \text{ plf})(3 \text{ in})}{2} = 97.5 \text{ in} \cdot \text{lb @ midheight}$$

Live loads:

$$P_{roof} = (15 \text{ ft})(20 \text{ psf}) = 300 \text{ plf}$$

$$P_{floor} = \left(\frac{13 \text{ ft}}{2}\right)(40 \text{ psf}) = 260 \text{ plf}$$

$$M_{floor} = \frac{260(3)}{2} = 390 \text{ in} \cdot \text{lb @ midheight}$$

Wind loads:

$$M_{wind} = \frac{25(9)^2}{8} = 253 \text{ ft} \cdot \text{lb} = 3038 \text{ in} \cdot \text{lb}$$

TABLE 10-25 Load Case 1 Calculations for Bottom Story in Structural Plain Concrete Example

$$U_1 = 1.4D + 1.7L$$

$$P_u = [1.4(300 + 65 + 1275) + 1.7(300 + 260)]$$

$$= 3248 \text{ lb @ bottom of wall}$$

$$M_u = 0$$

or

$$M_{u,\min} = P_u(0.10h) = 3248(0.10)(6) = 1949 \text{ lb}$$

$$\phi P_n = \phi 0.6 f_c' \left[1 - \left(\frac{l_c}{32h}\right)^2\right] A_1$$

$$= 0.65(0.6)(4000)\left[1 - \left(\frac{108}{32(6)}\right)^2\right] 72 = 76{,}781 \text{ lb}$$

$$\phi M_n = \phi 0.85 f_c' S = 0.65(0.85)(4000)(72) = 159{,}120 \text{ in} \cdot \text{lb}$$

$$5\phi\sqrt{f_c'} = 5(0.65)\sqrt{4000} = 205.5 \text{ psi}$$

Check:
$$\frac{P_u}{\phi P_n} + \frac{M_u}{\phi M_n} \leq 1$$

$$\frac{3248}{76{,}781} + \frac{1949}{159{,}120} = 0.055 \leq 1 \quad \text{OK}$$

Check:
$$\frac{M_u}{S} - \frac{P_u}{A_g} \leq 5\phi\sqrt{f_c'}$$

$$\frac{1949}{72} - \frac{3248}{72} = -18.0 \text{ psi} < 205.5 \quad \text{OK}$$

TABLE 10-26 Load Case 2 Calculations for Bottom Story in Structural Plain Concrete Example

$$U_2 = 0.75(1.4D + 1.7L + 1.7W)$$

$$P_u = [1.05(300 + 65 + 938) + 1.275(300 + 260)]$$

$$= 2082 \text{ lb}$$

$$M_u = [1.05(97.5) + 1.275(390) + 1.275(3038)] = 4473 \text{ in} \cdot \text{lb}$$

or

$$M_{u,\min} = 2082(0.6) = 1249 \text{ in} \cdot \text{lb}$$

Check:
$$\frac{P_u}{\phi P_n} + \frac{M_u}{\phi M_n} \leq 1$$

$$\frac{2082}{76,781} + \frac{4473}{159,120} = 0.055 \leq 1 \quad \text{OK}$$

Check:
$$\frac{M_u}{S} - \frac{P_u}{A_g} \leq 5\phi\sqrt{f_c'}$$

$$\frac{4473}{72} - \frac{2082}{72} = 33 \text{ psi} < 205.5 \quad \text{OK}$$

TABLE 10-27 Load Case 3 Calculations for Bottom Story in Structural Plain Concrete Example

$$U_3 = 0.9D + 1.3W$$

$$P_u = [0.9(300 + 65 + 938)] = 1173 \text{ lb}$$

$$M_u = [0.9(97.5) + 1.3(3038)] = 4037 \text{ in} \cdot \text{lb}$$

or

$$M_{u,\min} = 1173(0.6) = 704 \text{ in} \cdot \text{lb}$$

Check:
$$\frac{P_u}{\phi P_n} + \frac{M_u}{\phi M_n} \leq 1$$

$$\frac{1173}{76,781} + \frac{4037}{159,120} = 0.04 \leq 1 \quad \text{OK}$$

Check:
$$\frac{M_u}{S} - \frac{P_u}{A_g} \leq 5\phi\sqrt{f_c'}$$

$$\frac{4037}{72} - \frac{1173}{72} = 40 \text{ psi} < 205.5 \quad \text{OK}$$

References

American Concrete Institute, 1995
  Building Code Requirements for Structural Concrete, ACI Standard 318-95. Detroit, MI: American Concrete Institute.
American Concrete Institute, 1996
  ACI Manual of Concrete Practice, Part 3 - 1996. Detroit, MI: American Concrete Institute.
Ghosh, S. K., David A. Fanella, and Basile G. Rabbat, 1996
  Notes on ACI318-95: Building Code Requirements for Structural Concrete. Skokie, IL: Portland Cement Association.

Chapter

11

Energy Efficiency and HVAC

The energy efficiency of ICF walls is sharply higher than that of frame construction and results from three distinct mechanisms: high R-value, low air infiltration, and high thermal mass. Because of this, two important aspects of the HVAC engineering of an ICF home need to be handled differently. These are the sizing of the HVAC system and the consideration of air exchange. Doing either of these correctly depends on understanding those three mechanisms of energy efficiency and how they operate in an ICF home.

R-value

The most widely recognized attribute of a wall that influences space conditioning load is the R-value, or *thermal resistance*. The R-value of a plane of material or a sandwich of materials is its resistance to the conduction of heat from one side to the other. In the United States it is conventionally measured as h · sq ft · °F/Btu. The Fahrenheit temperature here is the temperature differential between the two sides of the plane, with heat flowing from the warmer to the cooler side.

Note that when dealing with assemblies of multiple materials, technical convention is sometimes to report the thermal transmittance, or U-value, rather than the R-value. U is measured in reciprocal units [Btu/(h · sq ft · °F)], so that a low U-value corresponds to a high R-value.

To avoid the cost of physical testing, engineers have developed many formulas and simulation algorithms for estimating the R-value of a material or wall. Different calculation methods are based on different assumptions and can give significantly different results. Unfortunately, this has led some building products suppliers to apply the method that assigns their material or system the highest R-value, seemingly regardless of whether the assumptions of that method fit the circumstances. A side effect, also unfortunate, is that the R-values reported for different products are often incomparable because they are based on different estimation methods.

Generally accepted as the most reliable method of determining wall thermal resistance is a physical test, the *guarded hot box* (ASTM, 1995). This procedure controls the temperatures on either side of the plane and measures the amount of heat energy that must be added to maintain the temperature on the warm side.

Two of the more sophisticated calculation methods, the parallel-path method and the isothermal planes method (ASHRAE, 1993), can be useful but require more careful interpretation. They attempt to approximate thermal resistance without testing. Depending on the type of wall system used, they will provide either under- or overestimates.

A few ICF systems have been tested with a guarded hot box. The completed walls, without finishes, are tested at an R-value of 25 to 35, depending on the system. Calculated R-values for the other ICFs suggest that all would test at an R-value of at least 18.

The R-value of a conventional frame wall is sensitive to the quality of construction and therefore varies. Calculation of the R-value of a 2×4 wood frame wall filled with average-density (R-11) insulation, without finishes, and of good construction shows it to be about 9.5, depending on the exact method of calculation used. Guarded hot box tests confirm an R of about 8 to 11, but with considerable variation (James and Goss, 1993). By extrapolation, with high-density (R-13) insulation the 2×4 wall's R-value should rise to an average of about 11. A 2×6 frame wall should test at about R-15 to R-16.

The R-values for frame walls are lower than the values of their insulation alone largely because of heat conducted through the lumber. Wood is significantly lower in R-value than insulation. The tested R-values of frame walls are often lower than the calculated R-values, largely because imperfections in construction leave various gaps in the wall without insulation, which are not recognized by the formulas.

Adding typical finishes will increase the R-value of either type of wall by 0.5 to 1.5.

Air Infiltration

The American Society of Heating, Refrigeration, and Air Conditioning Engineers (ASHRAE, 1993) estimates that 20 to 40 percent of the heating and cooling load in U.S. houses compensates not for heat conduction through the exterior envelope, but for air infiltration.

Frame walls have a large number of small gaps that allow air to pass through. Although air infiltration through an isolated section of an ICF wall has not been measured, it is generally believed to be negligible because of the "sealing" effect of the concrete pour and, secondarily, the low permeability of the foam facing on either side. This leaves the air infiltration through roof and fenestration unaffected, however.

Measured air infiltration rates in newly constructed frame houses average approximately 0.5 air changes per hour (ACH) at naturally occurring air pressures (ASHRAE, 1993). This means that a volume of air equal to half the amount already present in the building can be expected to enter it each hour. Tests of houses built with predominantly ICF exterior walls show these rates typically to range from 0.10 to 0.35 ACH. Some of the variation may result from how the walls are constructed, but much probably comes from other parts of the envelope (doors, windows, roof).

Thermal Mass

Recent engineering studies have confirmed what many ancient peoples knew: structures with exterior walls made of heavy materials enjoy more even interior temperatures and consume less conditioning energy than equivalently insulated structures made of light materials. The principle at work is informally called *thermal mass* or the *thermal flywheel*.

The uninsulated adobe homes of the Southwest provide one of the clearest intuitive examples of the thermal mass effect. During the hot daytime desert sun, the adobe walls warm only slowly. As a result, the interior remains cooler than it would be in a house with wooden walls. By the time the outdoor air cools in the evening, the walls have become warm. At night some of their stored heat passes to the interior, keeping it warmer than a house made of lighter materials. The cycle repeats day after day.

The magnitude of the thermal mass effect in a given building is directly related to (1) the heat capacity of the walls, (2) the temperature pattern of the local climate, and (3) the thermal resistance of the walls. The *heat capacity* of any material is the amount of heat energy required to raise the temperature of a unit volume of the material by a unit temperature, measured in the English system as Btu/(cu ft · °F). The heat capacity of a wall, as the term is commonly used, is the amount of heat required to raise a unit area of the wall by a unit temperature [Btu/(sq ft · °F)]. This clearly depends on what materials comprise the wall and how great a volume of them occur in each square foot. As outlined in Table 11-1, concrete has a greater heat capacity than common lumber. But in addition, each square foot of a typical ICF wall contains four times as much concrete (by volume) as a frame wall contains lumber. These factors combine to give typical unfinished ICF walls a heat capacity of about 12 Btu/(sq ft · °F) versus about 1.3 for 2×4 frame walls. Note that different types of ICF systems differ widely

TABLE 11-1 Heat Capacities of Alternative Wall Systems

Wall system	Identity	Predominant structural material		Wall heat capacity [Btu/(sq ft · °F)]	
		Heat capacity [Btu/(cu ft · °F)]	Density in wall (cu ft/sq ft)	Unfinished	With common finishes*
Wood frame					
2×4	Pine	18	0.073	1.3	2.6
2×6	Pine	18	0.115	2.1	3.4
ICF†					
Thin	Concrete	32	0.154	4.9	6.2
Average	Concrete	32	0.375	12.0	13.3
Thick	Concrete	32	0.500	16.0	17.3

*Assumed interior finish is ⅜-inch gypsum wallboard. For exterior, assumes heat capacity that is halfway between the highest-capacity common finish (½-inch portland cement stucco) and the lowest (vinyl).
†For comparative purposes, table assumes three different, but commonly used configurations of ICF walls. "Thick" assumes flat system with a maximum concrete thickness of 6 inches. (The widest forms commonly used have a maximum concrete thickness of 8 inches.) "Average" assumes an uninterrupted grid system with a maximum concrete thickness of 6 inches. "Thin" assumes post-and-beam wall with 5-inch-diameter members spaced every 2 feet vertically and 4 feet horizontally. (The greatest common spacing is 4 feet vertically and 8 feet horizontally.) See Chapter 2 for definitions of types of walls.

in their concrete contents and, therefore, their heat capacities. Also note that finishes add slightly to the heat capacities of all walls.

The critical aspect of the local climate is how frequently the outdoor temperature fluctuates about the building's interior HVAC set point. In theory, if the building is maintained at 70°F and the outside air fluctuates between 60 and 80°F over a 24-hour day, thermal mass will have its maximum effect, canceling out extremes of cool and warm periods. In times and places where the outdoor temperature is consistently below or consistently above 70°F the impact of mass will be less. In practice this means that the greatest HVAC savings from thermal mass occur along the southern border of the continental United States. In contrast, the effect is estimated to be sharply smaller along most of the northern border, in Canada, or in Hawaii.

The thermal resistance of the wall moderates the thermal mass effect because the effect depends on the conduction of heat into and out of a structure's walls. More of this conduction occurs when the R-value of the wall is low. Heavily insulated walls enjoy less incremental benefit from thermal mass.

Although calculation of the mass effect is complex, engineering simulations have made estimates for a variety of circumstances (Byrne and Ritschard, 1985; Wilcox, Gumerlock, Barnaby, Mitchell, and Huizenga, 1985). They compare, for example, the annual heating plus cooling load of two well-insulated (R-20 wall) houses, one with little heat capacity and one with a heat capacity of 10 Btu/(sq ft · °F). These equate approximately to a conventional frame wall that has been carefully superinsulated and an ICF wall with slightly less concrete than average and one of the lower ICF R-values. The simulations suggest that the wall with the higher thermal mass will, in the southernmost cities of the continental United States, consume 7 to 8 percent less heating and cooling energy than the light-frame house. This is so despite an assumption of no differences in the roofs or fenestration, and despite the assumed identical R-values of the walls. The savings in the northernmost cities were estimated at 2 to 4 percent, with a gradation in savings for cities between the South and the North.

Note that thermal mass savings have not been determined for ICF walls per se. Their design, which places insulation on both sides of the thermal mass, is unique and of uncertain impact. Actual savings may be different from the numbers presented here if ICF walls indeed behave substantially differently from other high-mass walls. Future experimentation may provide amended, more precise figures.

Combined Effects

The only available data that combine all factors into an estimate of total HVAC savings come from the real-world test: anecdotal reports of the utility bills of houses built with ICF walls, compared to bills of neighbor frame houses of similar design and time of construction. As discussed in Chapter 1, our own interviews with homeowners suggest that ICF home HVAC consumption ranges from 25 to 50 percent below that of frame houses of comparable size and design. Some sources claim even greater reductions.

Correction for thermal mass effect

There are some methods of combining two or more of the factors described in this chapter into a single number measuring the energy efficiency of a wall. One common approach is to compute an R-value that includes the effect of thermal mass. We can call such an R-value a *mass-corrected* R-*value,* although there is little consistency in how different parties refer to or describe it. The mass-corrected R-value for a wall of type X is the R-value that conventional frame walls would need to have to reduce the HVAC load of the frame house so that it equals the load of a house with type X walls (taking into account the R-value and thermal mass of the type X walls). Note that this method will not account for any savings in type X construction from lower air infiltration.

TABLE 11-2 Hypothetical Calculation of Mass-Corrected R-Value

	Conventional R-value (sq ft · °F · h/Btu)	Annual heat loss/gain (million Btu)	Mass-corrected annual HVAC load (million Btu)
Frame wall (standard 2×4):			13.0
Walls	9	5	
Roof	20	5	
Fenestration	4	3	
ICF wall:			9.0*
Walls	25	1.8	
Roof	20	5	
Fenestration	4	3	
Frame wall (superinsulated):			9.0
Walls	45	1.0	
Roof	20	5	
Fenestration	4	3	

*The mass-corrected load is less than the sum of the heat losses or gains from the separate envelope components because of an 8 percent reduction (assumed) from the thermal mass effect. Note, however, that in reality the HVAC load for the ICF house could be even lower because of additional savings from lowered air infiltration. These savings are not included in this example.

Table 11-2 provides a simplified numerical example. We assume a house of 2×4 construction has walls with an R-value (conventionally measured) of 9, and that the HVAC loads from conduction are as follows: through the walls 5 million Btu per year, through the roof 5 million Btu per year, and through doors and windows 3 million Btu per year. The total annual load is thus 13 million Btu. If the walls are instead built of an ICF with a conventional R-25 thermal resistance, the conduction through the walls is reduced by a factor of 9/25 to 1.8 million Btu per year. This would sum to a total load of 9.8 million Btu per year. However, the thermal mass effect reduces the load even further. We assume here that the house is located in a temperate climate, resulting in an incremental saving in total load of 8 percent. The load would then be 9.0 million Btu per year. To match this total load by adjusting the walls in the frame house, one would have to increase their R-value to 45. At this point they would allow only 1.0 million Btu in conduction and fully compensate for the lack of any benefit from thermal mass in the frame house. The mass-corrected R-value of the ICF walls is therefore 45.

Construction Technology Laboratories (CTL, 1996) has estimated the benefits of thermal mass to ICF buildings, using more sophisticated methods than those outlined here. Quoting from their 1996 report:

Insulating concrete form (ICF) walls have high thermal mass which evens out the effect of external temperature swings and allows ICFs to perform much better than wood frame walls. Construction Technology Laboratories in Skokie, Illinois, has predicted the R-value of a wood frame wall that would have the same total heating and cooling load as a typical 9 inch ICF wall in 38 U.S. cities. Using the ENVSTD compliance program for ASHRAE/IES 90.1-1989, *CTL has predicted that the* R-*value of a wood frame wall with equivalent performance would have to be greater than 50 in a majority of the cities analyzed.* The typical 9 inch ICF wall was assumed to have an R-value of 17.8 and a concrete thickness of 5 inches.

Two cautions are important when interpreting mass-corrected R-values. The first is that they must vary with the climate. Since energy savings from mass are greater in temperate climates, the proper correction depends on where the house in question is located. Thus a mass-corrected R-value can only be interpreted if one knows the city for which it was calculated. Note the careful reference in the CTL report to calculations being conducted for different cities, with some variation in the results. In other sources this sometimes goes unmentioned. The potential damage is (1) underestimating HVAC consumption by taking a mass-corrected value calculated with the climate of a southern city as representative of a system's energy efficiency in a northern area, or (2) overestimating consumption by taking a value calculated for a northern city and using it in a southern area.

The second caution is that the corrected R-value applies only to the walls. Building high-mass walls with a mass-corrected value of 45, instead of an R-9 for a 2×4 wall, does not mean that the entire house is five times as efficient and will therefore consume one-fifth as much energy. As we see in the example, the ICF house is an estimated 25 percent $[(12-9)/12 = 3/12 = 25\%]$ more efficient, as measured by reduction in total heating load. To state the significance of the mass more accurately, if one were varying only the walls, one could build with the ICF in question, or somehow construct a frame wall with an R-value that is a true R-45, to get the same total heating load.

Correction for air infiltration

Some engineers have also estimated corrected R-values that take account of energy savings from the wall system's thermal mass *and* its rate of air infiltration. These are calculated in a fashion analogous to the calculation of mass-corrected R-value, and are subject to analogous cautions of interpretation. For ICFs they will be even higher than the R-values corrected for mass only. However, currently these methods are not widely used or cited.

Code Treatment

Almost all ICF systems have R-values higher than required of walls in the most stringent U.S. residential building codes. So there is generally little need to take account of code wall R-value requirements in designing with ICFs.

But in rare instances the high energy efficiency of ICFs can be used to an indirect advantage. The model U.S. codes all permit trading off R-values among envelope components, and a growing number of state and local codes are, in turn, adopting these provisions. The rules provide that any one component of a building's envelope may have a lower R-value than normally required of it, if other components exceed their required values by sufficient margins. So, for example, a building with more window area or a lower-R roof than normally allowed would still meet the code if the walls were of sufficiently high R. The exact numerical requirements and tradeoffs differ across codes, but are specified in each one. Thus ICF walls can permit unusual building features in the exterior envelope (e.g., large expanses of glazing) that would otherwise, with a lower-R wall system, be disallowed.

In fact, many codes are assigning particularly high R-values to ICF walls because of thermal mass. The *Model Energy Code* (Council of American Building Officials, 1995) is currently adopted by all three of the major U.S. model codes for their energy requirements. It is gradually appearing in local codes as they adopt the latest versions of the three model codes. The *Model Energy Code* (MEC) contains three tables (502.1.a–c) that authorize specific increases in the required U-values (equivalent to decreases in the required R-values) of walls with heat capacities exceeding 6 Btu/(sq ft · °F). Under this criterion, most ICF walls would qualify (see Table 11-1 for data). If we look in the tables under 2000 to 4000 degree-days (corresponding to a moderate U.S. climate), we find that a wall normally required to have a U of 0.10 (that is, at least R-10) may have a U of up to 0.12 (R as low as 8.33) by virtue of its thermal mass. This is of little use in meeting the baseline R-value requirement for a wall, since the R-18 to R-35 of ICFs already far exceed nearly all local requirements. However, it could conceivably help achieve compliance with the energy portions of the code by compensating for other parts of the envelope that are below their baseline requirements.

Other codes perform the calculation differently, but permit similar U-value increases or R-value reductions. Their provisions usually include tables of U- or R-values, or "correction factors" to apply to the conventionally required U- or R-values. These values or factors are usually indexed by two factors. The first is the heat capacity or weight of the concrete in the wall. (Weight is sometimes used as an approximate measure of heat capacity.) The second is the number of degree-days of the local area, the popular measure of local climate patterns. The number appearing under the applicable parameters is the U or R that one may assign the wall for code purposes, or the amount by which one may adjust the conventional U or R.

Sizing Equipment

HVAC engineers currently use two methods for sizing equipment for ICF homes. First is the empirical method, which bases sizing on past experience with previous ICF houses. Second is calculation using conventional formulas or software tools, adjusted specifically to account for all three sources of energy savings: high R-value, low air infiltration, and high thermal mass. Under both methods, practice has understandably leaned toward conservatism.

Empirical method

The empirical method begins with a baseline range of energy consumption and adjusts it according to particulars of the specific house. There is no formal documentation or set of data for this rule-of-thumb technique. Based on interviews with HVAC engineers, a conservative baseline assumption would appear to be that an ICF house will consume 25 to 50 percent less heating and cooling energy than an otherwise identical house with 2×4 frame walls and fiberglass insulation.

The following is a list of the house characteristics according to which engineers typically adjust their estimates of energy load:

High conventional R-value of finished walls

Tight sealing around windows and doors and at roof

Temperate climate

High heat capacity of finished wall

When all are present, the tendency is to reduce equipment capacity by the maximum amount (50 percent). When none are present, reduction tends to be by the smaller amount (25 percent). When only some are present or some of them are only partially present, the reduction lies somewhere between the extremes. However, how to weight each factor is an individual decision made differently in practice by different HVAC professionals.

Adjusted calculation

Various formulas and software programs are available to calculate the energy load of a house. Those that include consideration of both the R-value of the walls and the air infiltration of the house lend themselves to producing useful load estimates for ICF homes. The procedure usually consists of inputting a mass-corrected R-value and an air change rate scaled to reflect the tightness of the house.

A few HVAC calculation tools provide means of entering thermal mass as a separate number. This is the technically preferred way to handle the calculation. When using tools that make no allowance for mass, one may instead simply use a mass-corrected R-value. Using a mass-corrected R-value takes account of both the high conventional R of the ICF walls and the contribution to energy efficiency from thermal mass. Thus both effects are combined into a single number.

However, the question remains of what mass-corrected R to use. The professionals we interviewed selected numbers ranging from 28 to 42, depending on whether they were located in the northern or southern United States. All felt that they were being conservative. The CTL study (1996) provides estimated mass-corrected R-values based on a representative ICF configuration for over 300 U.S. cities. These numbers are relatively scientifically justified, and are available to anyone who wishes to order the report from CTL or the Insulating Concrete Form Association (see Directory in Appendix C).

The correct air change rate to use again depends on judgment. Data indicate these rates are substantially below those of frame houses, but vary significantly. Presumably the variance is not a result of the ICF construction, but of the tightness of the window, door, and roof details. Active professionals approached this task also by an empirical method, assuming a low ACH for a house of tight details, and a higher one for looser construction. The data suggest assumed ACH values approaching 0.1 for a tight house and 0.35 for a loose one.

Ventilation provisions

There are few residential code requirements for supplemental ventilation, and there is diversity in the practice of HVAC engineers. Some engineers now routinely add fresh air intake to all houses, regardless of the construction. This makes the decision to ventilate automatic for their ICF projects. Some consider such provisions unnecessary even in ICF homes and do not particularly recommended them. They note that there have not been complaints among their clients with tight houses. Others make a judgment based on their projection of the air change rate of each house.

A popular conservative approach is to set up the HVAC and electrical systems so that it will be easy to add air intake equipment later if that proves desirable.

References

ASHRAE, 1993
1993 ASHRAE Handbook of Fundamentals. New York: American Society of Heating, Refrigeration, and Air Conditioning Engineers.

ASTM, 1995
"Standard Test Method for Steady-State Thermal Performance of Building Assemblies by Means of a Guarded Hot Box," ASTM Standard C236-89. In *1995 Annual Book of ASTM Standards*. Philadelphia, PA: American Society for Testing and Materials.

Byrne and Ritschard, 1985
Byrne, S. J., and R. L. Ritschard, "A Parametric Analysis of Thermal Mass in Residential Buildings." In *Thermal Performance of the Exterior Envelopes of Buildings III, Conference Proceedings*. Clearwater Beach, FL, December 2–5, pp. 1225–1240.

Council of American Building Officials, 1995
Model Energy Code. Falls Church, VA: Council of American Building Officials.

CTL, 1996
"Analysis to Determine Thermal Mass Performance of a Typical 9-in. ICFA Form Wall," Report to the Insulating Concrete Form Association, Construction Technology Laboratories, February 1.

James and Goss, 1993
James, Timothy B., and William P. Goss, *Heat Transmission Coefficients for Walls, Roofs, Ceilings, and Floors*. Atlanta, GA: American Society of Heating, Refrigeration, and Air-Conditioning Engineers.

Wilcox, Gumerlock, Barnaby, Mitchell, and Huizenga, 1985.
Wilcox, B., A. Gumerlock, C. Barnaby, R. Mitchell, and C. Huizenga, "The Effects of Thermal Mass Exterior Walls on Heating and Cooling Loads in Commercial Buildings." In *Thermal Performance of the Exterior Envelopes of Buildings III, Conference Proceedings*. Clearwater Beach, FL, December 2–5, pp. 1187–1224.

Chapter

12

Possible Future Developments

Many observers consider the methods used to engineer ICF structures overly conservative. The heart of the concern is that the ACI code is optimized for the large structures, high loadings, and regular, flat wall sections that dominate cast concrete construction in the United States. Several aspects of the ACI procedure are, some argue, inappropriate to low-rise residential ICF buildings. Specifically under question are:

1. The effective compressive strength of the concrete
2. Minimum reinforcement requirements

In response, the Portland Cement Association entered into a research program with Construction Technology Laboratories to build and test the structural behavior of walls constructed with a range of ICFs. This program is intended to provide empirical data useful to measuring the accuracy of conventional practice and suggesting potential refinements in engineering methods for the future.

Concrete Compressive Strength

Concrete poured into ICFs is believed to cure to a significantly higher strength than its allowed rating. This, however, is never taken into account in conventional engineering calculations.

The chart reproduced as Figure 12-1 shows how the compressive strength of concrete increases over time. It increases faster and reaches higher levels the longer it remains enclosed in its formwork. This occurs because the forms keep water in, extending the hydration reaction that hardens concrete. Since ICF formwork is ordinarily never stripped, concrete inside of it should achieve the maximum strength attained by concrete that is "moist cured the entire time" (see Figure 12-1). This is about 30 percent above the rated strength of

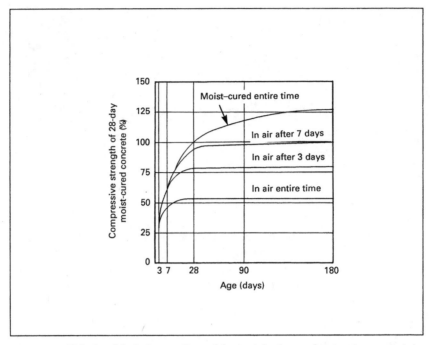

Figure 12-1 Relationship between time of form stripping and concrete compressive strength (*Portland Cement Association, 1994*).

the concrete, which by convention is the strength of unstripped concrete after 28 days.

Minimum Reinforcement

Engineers are also concerned that minimum reinforcement requirements are unnecessarily high for ICF walls. ACI specifies minimum reinforcement to bear shrinkage and temperature loads and general structural loads. Among other things, this requires separate reinforcing bars at least every 18 inches on center throughout the wall. The concern is that the ACI code is written to cover all cast-in-place concrete structures, including the very largest. Its minima might be appropriate for those extreme cases, yet greater than needed in the smaller, low-rise buildings that comprise the bulk of the ICF structures produced to date. Moreover, since the reinforcement of ICF walls is insulated from extreme temperatures, the temperature impact on the wall will be less than assumed in the determination of the minimum reinforcement.

As noted in preceding sections, ICF manufacturers generally recommend deviating from the ACI-required minima. The code allows this if complete calculations show that structural requirements call for less reinforcement. However, some feel that freedom from the standard minima should be a matter of course: for small residential buildings no justification should be necessary for the rebar schedule other than proper engineering for the endurance of the structural loads.

Ongoing Research

In late 1995, Construction Technology Laboratory (CTL) began an extended research program into the structural behavior of ICF walls under the direction of the Portland Cement Association (PCA). CTL is performing advanced calculations and physical testing on wall sections constructed with various flat and grid systems.

An immediate objective of the research is to determine whether walls built with reinforcing schedules that do not meet ACI minimum requirements do indeed exhibit predictable strength. Such a finding would give justification to the use of ACI 14.2.7, the provision that allows deviation from minimum reinforcement when structural analysis shows adequate strength and stability.

The program sponsors plan to produce several useful aids for the ICF engineer, including:

1. Design charts for various common wall cross sections relating steel reinforcing schedules to wall strength. These would be intended to facilitate quick design of ICF walls.
2. Prescriptive design tables of the reinforcement required for walls meeting limiting criteria such as limited loading, minimum thickness, maximum height, and so on. Diagrams and wall section details would accompany these. They would provide quick guidance on the design of the types of walls most commonly encountered in single-family home construction.
3. PCA-WALL, a computer program that performs standard engineering calculations on ICF and other slender walls.

The computer program and research report are available from the Portland Cement Association. (See the Directory of Product and Information Sources in Appendix C.)

Use of Structural Plain Concrete Provisions

According to the directors of the PCA-CTL research program, one of the outcomes of their work may be a recommendation that many ICF houses be designed in accordance with Chapter 22 ("Structural Plain Concrete") of the ACI building code. This section achieved chapter status with the 1995 edition of the code.

In contrast to the code sections currently used for ICF engineering, Chapter 22 takes fuller account of the strength of the concrete itself in calculating total wall and lintel capacity. Therefore steel requirements are frequently less.

The structural plain concrete provisions promise to fit ICF walls and low-rise residential construction better than alternative sections of the code. Minimum reinforcement requirements, thought by many to be excessive in these circumstances, are automatically waived when designing to the provisions of the chapter. When engineering under these guidelines, it should be more often apparent where the light loads of residential construction are met by the strength of the concrete alone. Lastly, by considering nearly the full concrete strength of all structural members, this procedure promises to value the contributions of unreinforced members and wall sections more accurately.

There are a few uncertainties in the application of the method of structural plain concrete. This method is not as familiar to many engineers as other sections of ACI 318. Moreover, it relies more heavily on proper field application to achieve the full rated strength of the concrete. This would include, for example, thorough consolidation, which can be slightly trickier with ICFs because of its potential to damage the formwork (see "Consolidation" in Chapter 15).

Nonetheless, future research may reveal that the structural plain concrete method, with a few caveats, is a more appropriate engineering paradigm for ICF house design.

Advanced HVAC Calculation

It is likely that new summary measures of wall energy efficiency will appear in the near future. Oak Ridge National Laboratory is currently developing a rating system that will include the effects of R-value, thermal mass, and air infiltration, as well as moisture effects (Christian and Kosny, 1995). If such efforts are successful, they would allow convenient summary comparison of alternative wall systems and methods of construction. They might also simplify the task of sizing HVAC equipment.

References

Christian and Kosny, 1995
    Christian, Jeffrey E., and Jan Kosny, "Toward a National Opaque Wall Rating

Label." In *Thermal Performance of the Exterior Envelopes of Buildings VI, Conference Proceedings.* Clearwater Beach, FL: December 4–8, pp. 221–240.

Portland Cement Association, 1994
Design and Control of Concrete Mixtures, 13th ed. Skokie, IL: Portland Cement Association.

Part

5

Assembly

There are alternative ways to perform many of the steps of ICF construction. Knowing these is particularly valuable for the builder and trades. However, understanding the assembly process also provides useful background for the construction manager and for the designer interested in specifying procedure at an especially fine level.

Note that this part presents only generalizations on typical ICF assembly. Each ICF system is proprietary and unique, and the manufacturers have unique specifications for proper use of their systems. Moreover, these specifications can change over time. **The current specifications and directions of the ICF manufacturer always take precedence over the information presented in this book.**

Chapter

13

Process Overview

The first provision made in construction for an ICF wall is usually the installation of protruding rebar (also called *dowels*) in the foundation below (Figure 13-1). This is true whether the wall rests on a slab, footing, or conventional concrete wall foundation (basement or stem wall).

Once the base and dowels are in place, typically the crew places the first course of ICF units around the perimeter. In the case of the largest units (panels 8 to 12 feet tall), this may complete the first story (Figure 13-2). For all other units, multiple courses are necessary, as seen in Figure 13-3. The crew levels the first course carefully.

Figure 13-1 Dowels protruding from a slab foundation. (*Amhome U.S.A. Inc.*)

262 Assembly

Figure 13-2 Setting 8-foot-tall panels to form first-story walls. (*Amhome USA Inc.*)

Various methods are available to accomplish this, but for most systems it is critical. An uneven first course can require time-consuming correction in the courses above.

Most often the crew completes each course in its entirety around the perimeter before beginning the next. This contrasts with concrete masonry practice, in which corners may be built up several courses in advance of the center of the wall. Unless the perimeter is strictly dimensioned (see "Dimensioning" in Chapter 6), on each course one unit along each wall must be cut to achieve the precise design length of the wall. Figure 13-4 provides an example.

A strict rule of assembly is that both vertical and horizontal cavities must remain in alignment. Misaligning grid or post-and-beam units would leave narrow points in the final concrete members, weakening the structure. Misaligning any system also throws the fastening surfaces (if any) out of alignment. This sharply increases the time and difficulty of attaching many exterior and interior finishes. Figure 13-5 shows correctly aligned cavities.

Fortunately, consistent alignment is not difficult. It is assured horizontally as long as courses are kept level. Post-and-beam blocks and

Process Overview 263

Figure 13-3 Setting medium-sized units in multiple courses. (*Lite-Form Inc.*)

Figure 13-4 Cutting the end of a block unit to achieve proper length.

264 Assembly

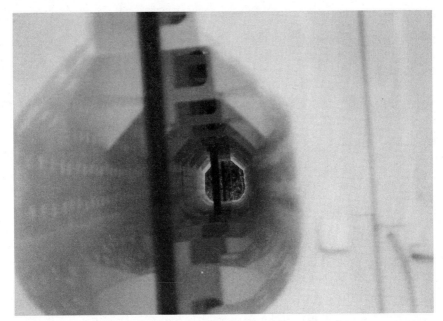

Figure 13-5 Aligned horizontal cavities. (*New England Foam Form.*)

some grid blocks have interconnects that force vertical alignment, and in most other systems the task is simple and well stressed in the documentation. There is one key practice for maintaining vertical cavity alignment: the cuts made on each course to bring the wall to correct length need to be lined up as closely as possible.

Some other important tasks must occur as the wall is stacked. The crew inserts the termite shield (if any) when the wall is at the appropriate height. The crew cuts units as necessary to leave precise openings for windows and doors. There are multiple methods of accomplishing this (see "Opening Formation" in Chapter 14). Most include insertion of a wooden or plastic buck to frame the perimeter of the opening (Figure 13-6). In addition, every time the form wall reaches a height at which horizontal rebar is specified, the crew sets that in place before proceeding to the next course (see Figure 13-7).

When the first few feet of formwork is set, the crew usually puts bracing or scaffolding in position. Bracing has two purposes. It holds the formwork precisely plumb and square, and it is sometimes used to reinforce potential weak spots. However, the latter can also be achieved with alternatives such as glue, tape, or pieces of strapping or plywood fastened to the wall. A common bracing consists of vertical studs (sometimes called *strongbacks*) against the wall every few feet, supported by a diagonal strut (called a *kicker*) to the ground (Figure 13-8). In addition, it is common practice to run a line of 2×4 lumber along

Process Overview 265

Figure 13-6 Inserting wooden buck into a notch in form wall, cut to proper opening position.

Figure 13-7 Setting horizontal rebar. (*Quad-Lock Building Systems.*)

Figure 13-8 Form wall braced with strongbacks and kickers. (*3-10 Insulated Forms L.P.*)

either side of the top of the formwork around the entire perimeter. The lumber might be fastened directly to the formwork (Figure 13-9), or the inside and outside rails might be connected by strapping to form a "ladder" resting on top. This *top bracing* is designed to reinforce the formwork along the top edge. Some crews will omit the strongbacks and run occasional kickers directly to the top bracing.

Various types of scaffolding are adequate. Beyond certain elevations, however, OSHA requires that it have toe boards and hand rails (see Chapter 2). Scaffolding elevates workers to set the top courses of units and place concrete. Bracing and scaffolding systems available from ICF manufacturers combine the two functions. Vertical studs or posts both brace the wall and support scaffold planks along it. Figure 13-10 provides an example.

Several steps occur between setting the top course of a story and placing the concrete. The crew drops vertical rebar into position. They insert joists or connectors into the formwork to provide for the floor deck that will come on top of the wall (as seen in Figure 13-11), or, if this is the top story, they attach connectors for the roof. They place sleeves of plastic pipe into the forms to create penetrations through the wall as necessary for plumbing, HVAC, and electrical lines. Figure 13-12 pictures an example. They double-check all assembly and bracing and look for potential weak spots.

Process Overview 267

Figure 13-9 Attaching 2×4 along top edge of a form wall as top bracing. (*Portland Cement Association.*)

Figure 13-10 Combined bracing and scaffolding. (*Lite-Form Inc.*)

Figure 13-11 Floor joists attached with special hanger set into wall. (*3-10 Insulated Forms L.P.*)

Figure 13-12 PVC pipe installed in formwork as sleeve for utility lines that will penetrate final wall.

Process Overview 269

Figure 13-13 Placing concrete through sill of a buck. (*Reddi-Form Inc.*)

During the pour the crew places concrete into the formwork, as illustrated in the Introduction (Figure I-4). They usually do this in a series of lifts of 2 to 6 feet to avoid placing high pressures on the forms. Some concrete is placed through the gap in the sills of window bucks to fill the wall below (Figure 13-13). The crew also vibrates the formwork or the concrete directly to consolidate the concrete.

In rare instances a spot in the formwork ruptures during the pour ("blows out") and concrete spills out. This is generally the result of placing concrete too quickly or failing to reinforce weak spots. When a blowout does occur, it can be repaired faster than with conventional formwork (in 10 to 15 minutes) and without stopping the pour. The crew members placing concrete move temporarily along the wall while other workers empty excess concrete from the hole, then replace and reinforce the broken foam.

Concrete placement stops at or just below the top of the formwork. After the concrete has cured, the crew dismantles scaffolding and bracing for reuse. If this is the top story, they begin building the roof. If it is not, they complete the floor deck and work from it to stack the next level of formwork.

The construction of the upper stories closely follows the construction pattern of the first. Upper-story units are placed directly on top of those below.

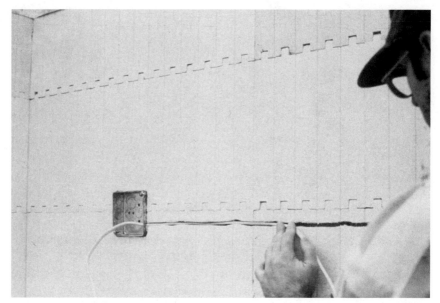

Figure 13-14 Placing electrical cable into a channel cut into foam surface. (*Portland Cement Association.*)

After the walls are complete, the building sequence is conventional, with two major exceptions. First, because the ICF walls are tight to the weather, they need not be sided before interior work begins. Second, they need not be insulated.

Inside work begins with attaching abutting interior walls to the exterior ICF walls. When the time comes to run utility lines (primarily electrical cable and piping), the trades cut channels in the foam of the ICF walls and push the lines into place. See Figure 13-14 for an example. Conventional interior finishes are then attached to the surface. Small changes in the finish application method (described in Chapter 7) are sometimes involved.

Outside, any of the conventional exterior finishes are also attached in the conventional manner, with a few minor adjustments (as described in Chapter 7).

Chapter 14

Formwork

The most novel aspect of wall assembly is the creation of formwork from the ICF units.

Crew Composition

Effectively constructing ICF walls requires both carpentry and concrete skills. The precision of carpenters is important for constructing accurate, square bucks, cutting clean openings, and building level, plumb walls. Setting rebar and specifying and placing concrete are important skills most familiar to concrete forms crews.

There are three commonly used methods of obtaining a crew with the complete set of necessary skills. Table 14-1 outlines them.

A rough carpentry (framing) crew can receive instruction in ICFs, including concrete skills. Many framing crews have formed footers or other small pours before. However, if their familiarity with concrete is particularly weak, the crew might also be supplemented with an additional person.

An advantage of using such a crew is that the precision activities (cutting to dimension, leveling) tend to be well executed. In addition,

TABLE 14-1 Wall Crew Alternatives

Wall crew	Limitations	Advantages	Disadvantages
Carpentry	None	Precision skills, covers framing	Less concrete skill
Forms	None	Concrete skills	Less precision skill
Custom	None	Includes all skills	Logistics of forming a new crew, uncertain availability, potential cost

there will be no need to change crews when work shifts to roof and interior framing. At the extreme this one crew can form the footing, build the foundation walls, build all above-grade exterior walls, then build the roof and interior walls. This saves one crew as compared with conventional construction: the forms or masonry crew (for the foundation walls) and the rough carpentry crew become one. It also eliminates coordination problems that might arise between separate crews. The major disadvantage of using a carpentry crew is that care must be taken to ensure that concrete-related tasks are well executed, at least until this group is experienced.

Similarly, it is possible to train a concrete forming crew in ICFs and basic rough carpentry, or to supplement the crew with a skilled framer. The forming crew then takes responsibility for the above-grade exterior walls away from the framing crew. The framers construct the floors, roof, and interior. The advantage is that rebar and concrete placement is likely to be performed well. The disadvantage is that more care is necessary to ensure plumb, level, and square work.

It is also possible to assemble a new crew customized for the task. It would include about half carpentry workers and half concrete workers. The advantage of a custom crew is that it includes the optimum mix of personnel to build an ICF wall. Whether it can also handle other tasks efficiently (such as framing the roof and interior walls) depends on the workers' skills. The disadvantages are the same as for forming any special-purpose crew. Such crews are not already available; the project management must create them. The time involved is likely to be paid back in greater efficiencies only if the crew can be kept intact and active for extended periods. In addition, if they are underutilized the individuals will likely leave for steadier work in more widely needed trades. On balance, custom crews are most attractive when a large project or series of smaller ones will keep them busy for an extended period.

Dowel Placement

The dowels protruding from a non-ICF foundation (concrete footing or wall) need to be positioned accurately so that they are in the center of the ICF wall above. In grid and post-and-beam walls they must also be in the center of the correct vertical cavity. However, efficiency dictates that the placement be done quickly, too.

Table 14-2 contains the popular alternatives for dowel placement. The first is to position them in advance of pouring the footing or wall, holding them in place with wire. This requires calculating the positions in which ICF cavities will fall precisely and in advance. Error can result either from miscalculation or because the actual layout of

TABLE 14-2 Dowel Placement Alternatives

Dowel placement	Limitations	Advantages	Disadvantages
Prepour	None	Familiarity	Difficulty of precise positioning
During pour	None	Speed	Time pressure, difficulty of precise positioning
Drilled	No angled dowel	Precise positioning, ease of adjustment	Drilling required, material requirements, potential weakness

formwork varies from how it is projected on paper. However, it is a procedure familiar to concrete crews.

It is also possible to push the dowels into position in the wet concrete just after it has set enough to hold them upright. This requires the same advance calculations and adds some time pressure. However, it is fast because no wiring is necessary.

The third alternative is to delay rebar placement until the first course of formwork is set on the concrete below. Since the concrete will be hard, this requires drilling into it to make holes for the bar. It also requires an appropriate adhesive, such as a concrete epoxy, to fasten the bar securely. And it is possible only with a straight dowel. This eliminates the possibility of an L dowel, which some designers specify because of its greater pullout strength. But positioning can be extremely precise. Since the formwork is in place, locating an accurate center is simple.

Naturally, one can place dowels according to one of the first two methods and, if errors result, later cut off the affected bars and drill them into their proper positions.

Unit Orientation

When using most panel or plank systems it is possible to stack the units horizontally or vertically, as shown in Figure 14-1. Table 14-3 contains the considerations. Forming curved walls is usually a little easier with horizontal orientation because it allows arcing a unit along its longest dimension. (See "Curved Walls" later in this chapter.) This gives one leverage in bending the foam. Horizontal stacking also requires less cutting when forming a narrow horizontal wall section, such as the section between two closely spaced windows, one above the other.

Conversely, if one wishes to form a series of short, angled wall sections or a narrow vertical column (as between two closely spaced side-

274 Assembly

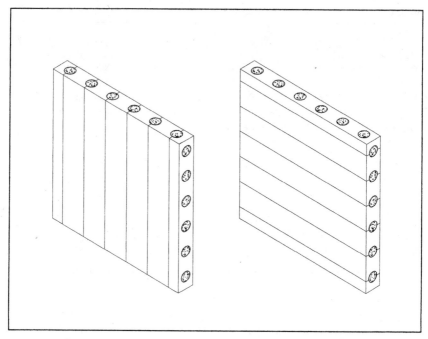

Figure 14-1 Orienting panel units vertically (left) and horizontally (right).

TABLE 14-3 Unit Orientation Alternatives

Unit orientation	Limitations	Advantages	Disadvantages
Horizontal	None	Ease of curving, ease of forming beams	Difficulty of forming columns
Vertical	Certain non-block systems	Ease of faceting planks, ease of forming columns	Height of course, difficulty of forming curves

by-side openings), vertical orientation can save time and cutting. With 8-foot-long units it has the disadvantage of placing one end above eye level immediately. This can be awkward for precise work unless scaffolding is already in place.

It is also possible to switch between horizontal and vertical stacking on the same wall. Frequently when using planks or narrow panels, crews will stack horizontally until they encounter narrow vertical wall sections. For these they will switch to vertical stacking, then switch back again. Some care is necessary here. Both vertical and horizontal cavities must be kept aligned, and connections between oppositely oriented units may need reinforcement.

Formwork

TABLE 14-4 Course Leveling Alternatives

Course leveling	Limitations	Advantages	Disadvantages
Dry set	None	Unlimited time, ease of changes	Labor, need for guides
Wet set	Wide base	Combines setting and leveling, tight base joint	Difficulty of changes, time pressure, floating of units

Course Leveling

Unless the footer or foundation below is precisely level, it is necessary to bring the first course of formwork to level. There are two major methods of doing this. Table 14-4 notes considerations in using each.

The first is the *dry set*. Under this method one waits for the concrete below to harden, then shim or shave the units as necessary to bring their top edges precisely to a level string line. Gaps between shimmed blocks and the footer may be filled with foam adhesive. Figure 14-2 depicts the principle, and Figure 14-3 shows it in practice. Under this method, the crew usually begins by putting chalk

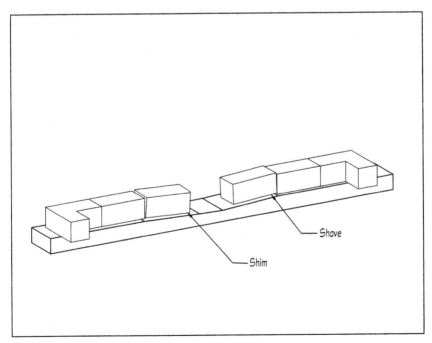

Figure 14-2 Points requiring shimming and shaving of first-course units to conform to an uneven footer.

Figure 14-3 Shimming at a low spot in footer. (*AFM Corporation.*)

lines on the foundation to mark the intended inside and outside surfaces of the form wall. They then install two continuous lines of 2× lumber or metal or plastic channel (known collectively as *guides*). The first course of units goes between the guides. This arrangement braces the bottoms of the units against the force of the concrete during the pour.

The advantage of the shim and shave technique is that the crews can take their time double-checking the layout, and changes are easy to make before the second course goes up.

The alternative, called *wet setting,* is possible when the foundation below will be thicker than the width of the form units. In practice this means it is generally limited to projects in which the ICF walls begin on a footing. Under this method the crew positions a string line at the intended top edge of the first course before the footing is filled with concrete. Immediately after concrete placement, they set the first course, pressing each unit into the wet concrete $\frac{1}{4}$ to $\frac{3}{4}$ inch to bring its top edge to the string. Figure 14-4 shows a wet set under way.

In the wet set method, the hardened concrete will hold the bottoms of the units securely, eliminating the need for guides. The work of shimming and shaving is also avoided. It does require quick work, and correcting errors after the concrete is set can be messy and time-consuming. In addition, some contractors have complained that the

Formwork

Figure 14-4 Wet setting the first course of units. (*3-10 Insulated Forms L.P.*)

foam units can float out of position. However, those experienced with the method claim that the difficulties disappear with practice.

Note that the wet set method is best suited to planks and blocks that are oriented horizontally. With tall units one must handle most tasks that are concurrent with stacking (such as making cutouts for openings, setting rebar) while setting the first course. The time in which concrete sets does not allow for this.

Set Sequence

A course of ICF units can be set in a variety of sequences (listed in Table 14-5 and depicted in Figures 14-5 through 14-7). One is to set all corner units, then work inward to some intermediate point along each

TABLE 14-5 Set Sequence Alternatives

Set sequence	Limitations	Advantages	Disadvantages
Corner-in	None	Can minimize cuts	Potential cut misalignment
Corner-corner	None	Aligns cuts, cuts at corner	Extra cuts
Perimeter	None	Can minimize cuts	Cuts in corner units

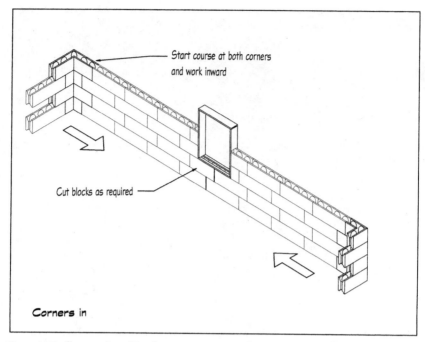

Figure 14-5 Corners-in setting leaves cuts near wall center and around openings.

wall ("corner-in"). The major advantage is that the point at which units from each direction meet can be set so as to minimize cutting. Optimal is to stack inward toward a door opening. Units will have to be cut along either side of the opening (as always), but not again to reach accurate wall length until stacking proceeds above the opening.

Such corner-in stacking is most advantageous when using units of low and moderate heights (blocks and planks). With these systems the reduction in cutting is greatest. Since this method does not dictate a single cut point, the stackers must exercise some care to make sure that vertical cavities remain aligned.

An alternative is to set the corners, then stack in a standard direction (for most crews, from left to right) from one corner. This method is called "corner-corner." One makes any necessary cuts on the last unit before the ending (usually the right) corner. This places all cuts in a consistent position, minimizing the potential for misalignment. In addition, there is typically bracing at the corner anyway, and that can often be used in reinforcing the nearby cuts, if necessary.

The major disadvantage is extra cutting. Each course requires a cut to achieve the correct wall length, in addition to those made alongside openings.

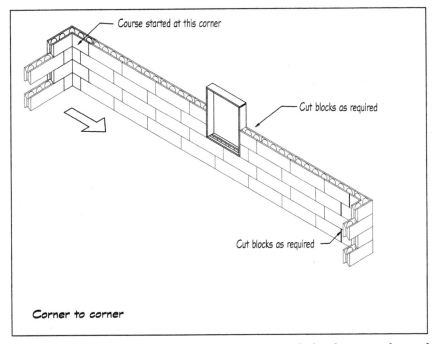

Figure 14-6 Corner-corner setting leaves cuts adjacent to right-hand corner and around openings.

Perimeter setting starts at one corner and goes around in one large circle. Corner units are stacked in sequence, just like stretchers. The cuts to achieve wall length are typically made at the corner. This method is popular with large user-assembled units. These are usually cut at the corner anyway to achieve the correct offset between inside and outside face shells. Thus that function and the achievement of wall length can be handled in one cut.

Bond Pattern

As with masonry, it is possible to offset successive courses of ICF units in different ways (see Table 14-6 and Figure 14-8). Preferred with most shorter units (4 feet and below) is a conventional running bond. This staggers the vertical joints between adjacent courses, usually by one-half the length of a unit. This is also the most common pattern in masonry, and it is adopted for similar reasons here. Aligned vertical joints form a weak spot during the concrete pour, and this pattern avoids them. It does, however, require keeping track of the correct break points.

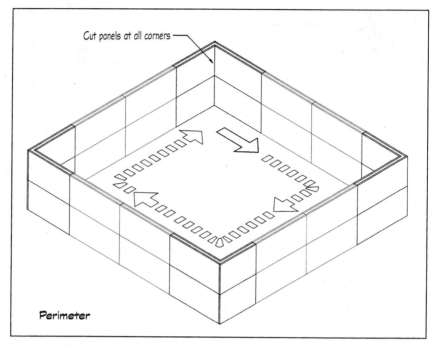

Figure 14-7 Perimeter setting leaves cuts in corner units.

TABLE 14-6 Bond Pattern Alternatives

Bond pattern	Limitations	Advantages	Disadvantages
Running	None	Lack of weak points	Moderate complexity
Stack	None	Simplicity	Potential weak points, need for support
Double running	Field-assembled units	Lack of weak points	Complexity

The traditional stack bond aligns all joints up the entire height of the wall. This is often used with long (over 4-foot) plank systems, and usually adopted when using panels. With these systems there are fewer vertical joints, so reinforcing weak spots requires less effort. The method has the general advantage of simplifying the setting procedure: each course is (in this regard) identical to the one below.

One manufacturer (Polycrete) recommends a double running bond. Under this method the inside face shells are offset from the outside, in addition to the offset between courses of a running bond. It is only

Formwork

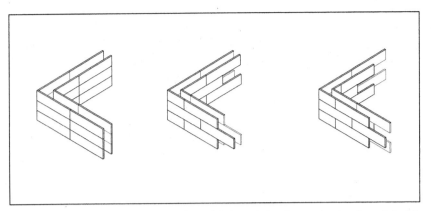

Figure 14-8 Setting form units in (left to right) stack bond, running bond, and double running bond.

TABLE 14-7 Interconnection Alternatives

Interconnection	Limitations	Advantages	Disadvantages
Standard interconnect	High-grip systems	Speed, low-cost	Potential weak spots
Weight	None	Speed	Potential unevenness, awkwardness during pour
Glues	None	Continuity of seal, high strength	Time, cost
Tape	None	Moderate time, moderate cost	Moderate strength, potential need for removal
Wire	Systems with exposed ties	Speed, strength	Need for removal

possible with user-assembled units. Its advantage is a reduction in potential weak spots, and its disadvantage is procedural complexity.

Interconnection

Holding the units together securely during the concrete pour is important. Otherwise they can shift out of position, taking the wall out of plumb and allowing concrete to leak. Several alternatives are available for interconnection (Table 14-7). Figure 14-9 depicts some of them.

One method is to rely solely on the standard interconnect of the system. The interconnects of some ICFs are designed to withstand the side and lift forces of the concrete. The manufacturers' documents cover this.

282 Assembly

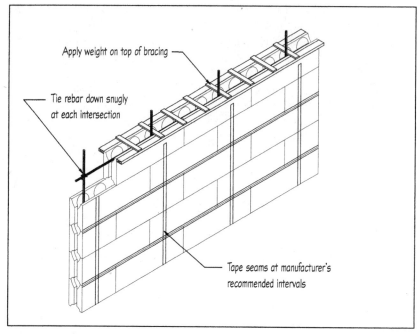

Figure 14-9 Some methods for reinforcing the interconnection of ICF units.

Included are most flat panel and flat plank systems and most blocks with interconnecting teeth. This is the fastest and lowest-cost alternative when it applies. Some installers have had concerns that the interconnects might not hold and have taken extra steps anyway. However, such precautions should become less important as the installer gains skill with a particular brand and the placement of concrete.

A second alternative is to rely on weight applied to the top of the wall. Such objects as concrete blocks or heavy scrap lumber are easily placed on top to hold the wall against motion. The disadvantages are: (1) it can compress the foam and make the wall slightly uneven, and (2) working around these objects can be awkward during the pour.

Gluing the joints is another option. Some crews run a bead of adhesive foam or construction adhesive along the edges of each unit just before placing it. Others apply construction adhesive across the surface of all joints after the formwork is up. This makes for a strong connection, and it also provides an air and moisture seal along each joint. However, it requires more labor and material than the alternatives.

It is also possible to tape the blocks together after they are in place. Most popular is common duct tape. It need not be applied along every (or even any) joint. One places vertical and horizontal lines of tape a few feet on center, forming a grid pattern on either side of the wall. This can require less time and resources than glues. Some are concerned it might not be as strong, and depending on the intended finishes, one may need to remove the tape.

When using a system with protruding ties (which includes most flat panel and plank systems), one can also loop wire around ties of adjacent panels, crossing over the joint (shown previously in Figure 2-25). This is popular because of the speed and strength of the connection. It is generally only practiced on long vertical joints because the horizontal and short vertical joints of the systems involved are usually adequately held by the standard interconnects. One may need to remove the wire before finishing.

Bracing

Various levels of bracing are possible (Table 14-8). The quickest and simplest is to rely solely on top braces and, connected to these, kickers. This is generally sufficient to plumb the wall. It will not prevent bulges that may occur. However, many experienced contractors claim careful pours prevent bulges. If these do occur, the final wall can be made straight again by shaving foam off the high spots, at some labor cost.

One step up is to include strongbacks periodically (usually every 4 feet) along both inside and outside of each wall. (Refer to Figure 13-8 for a photograph.) This adds some security against bulging, but it also adds time and labor. One must also have additional lumber on site, although it can be reused later.

The combined bracing and scaffolding systems (see Figure 13-10) available from many manufacturers provide the same bracing advantages as a lumber setup with strongbacks: plumbing of the wall and prevention of bulging. Some have turnbuckles that allow for fine adjustment of plumb even after concrete placement. Since they also provide

TABLE 14-8 Bracing Alternatives

Bracing	Limitations	Advantages	Disadvantages
Kickers	None	Speed, low cost, prevention of leaning	Potential for bulges
Strongbacks	None	Prevention of leaning, prevention of bulging	Time, labor cost, materials
Brace and scaffold system	Available systems	Prevention of leaning, prevention of bulging, combines two functions, easy adjustability	Unfamiliarity, initial cost

TABLE 14-9 Scaffolding Alternatives

Scaffolding	Limitations	Advantages	Disadvantages
None	None	Low setup cost, lack of setup time	Inefficiency
Conventional	None	Familiarity, efficiency at heights	Setup time, initial cost
Brace and scaffold system	Available systems	Combination with bracing, efficiency at heights	Unfamiliarity, initial cost

scaffolding, the setup time of the two are consolidated. They will require some learning by crews, and do involve an initial cost to purchase the equipment. Some manufacturers have lending and rental policies.

Scaffolding Alternatives

The primary purpose of scaffolding is to position workers for safe, efficient stacking of units when the form wall becomes high, and for concrete placement at the top of the wall. Alternative arrangements are listed in Table 14-9.

It is possible to rely on ladders, steps, or wheeled platforms instead of permanent scaffolding. Setup will be fast and inexpensive. However, the need to move equipment frequently can make this inefficient. It becomes somewhat more attractive when one uses a floor construction method that allows building the floor before placing concrete (see "Floor Joist Pocketing" in Chapter 16). In that situation it is possible to use the floor while placing the concrete.

Conventional scaffolding (Figure 14-10) permits more efficient work and is familiar to many crews, but initial setup is necessary. As noted previously, the combined bracing and scaffolding systems consolidate that setup with the installation of bracing. They require some learning and initial investment, however.

Regardless of the form of scaffolding you use, above a platform height of 10 feet, and sometimes below, OSHA regulations require hand rails and toe boards on all open sides and ends.

Curved Walls

There are three common methods of creating a curved form wall (see Table 14-10 and Figures 14-11 through 14-13). One is the method of bending units. This involves cutting slots from the inside face shell of each unit, then bending and gluing the unit back together. Creating these units requires specifications from the manufacturer or some

Formwork   285

Figure 14-10 Some conventional scaffolding. (*Lite-Form Inc.*)

TABLE 14-10 Curved Wall Alternatives

Curving method	Limitations	Advantages	Disadvantages
Bending units	Pure forms	Simplicity, single-piece unit	Unusual dimensions, potential difficulty of bending outside face shell
Bending face shells	Field-assembled units	Potential simplicity, ease of bending	Difficulty of aligning ties, many cuts
Angling units	Stack bonding	Familiarity, no bending	Time, labor, faceted surface

trial and error to determine how wide a slot to cut for a particular radius. However, it is relatively fast and simple and keeps units in one piece.

In practice, when using this method the crews may first assemble the section of the wall that will be curved to form a sort of wall panel. They then make the cuts up the entire length of the panel, rather than cutting each unit individually. This requires fewer cuts, and keeps the cuts precisely aligned.

With field-assembled units it is also possible to make regular dado cuts inside each face shell, bend and glue the shells, and only then assemble the two bent shells. This is fairly simple with flat plank systems. The dado procedure adds flex to the foam and reduces the risk

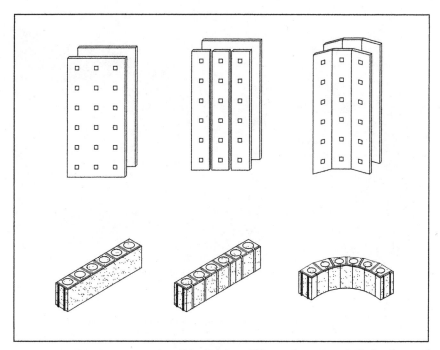

Figure 14-11 Forming a curved wall by bending ICF panels (top) and blocks (bottom). Frequently planks and blocks are prestacked into panel-size sections before cutting to form curves by this method (see text).

of breaking. The process involves more cuts than needed to bend units. Aligning ties when finally assembling the face shells can also be tricky.

The third option is to cut the ends of the units at angles and stack bond them. This actually creates a series of short straight wall sections, rather than a true curve. The exterior can be returned to a true curve after the pour, however, by sanding or shaving the foam to shape or by finishing with a material that can perfect the arc (such as thick stucco). Crews may be familiar with this general approach from lumber or masonry, and no foam must be bent. With many systems, however, it requires more time and labor. In addition to the cutting, the joints must usually be connected and reinforced well because they align vertically and because the standard interconnects are compromised. Achieving a true curve (if desired) also requires more work.

Long plank systems lend themselves well to forming curves through angling. Planks can be oriented vertically, so their vertical joints will be close together anyway. And some have interconnects that remain largely intact when the edges of the face shells are shaved.

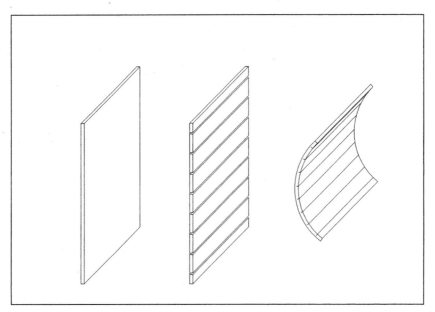

Figure 14-12 Forming a curved wall by bending face shells of user-assembled unit.

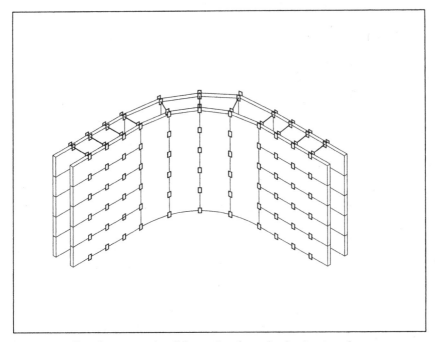

Figure 14-13 Forming a curved wall by cutting the ends of units at angles.

TABLE 14-11 Irregular Angle Alternatives

Irregular angle	Limitations	Advantages	Disadvantages
Field cut	None	Angle flexibility, no specialty units	Time and labor, potential variability
Precut or preformed	Standard angles, certain brands	Accuracy, time and labor savings	Order and inventory costs, materials cost
Hinge	Certain brands	Angle flexibility, time and labor savings	Order and inventory costs, materials cost

Irregular Angles

There are three common methods of forming irregular (non-90°) wall angles (Table 14-11). Miter cutting in the field to form corner units of the appropriate angle has the advantages and disadvantages of field cutting units in general: any angle is possible and no special units need be ordered, but field time and labor are greater, and there is some possibility that the task will be done in an inconsistent manner.

Precut or preformed specialty units (as shown previously in Figure 2-19) reverse these advantages and disadvantages. They are available only from some manufacturers, and usually only in one or two standard angles (usually 135 and 120°).

The pluses and minuses of hinge units (see Figure 2-20) are the same as those of the other specialty units, except that their angles can be adjusted through a wide range in the field.

Opening Formation

When formwork reaches the planned location of an opening, there are two common ways to form it. These are listed in Table 14-12. The first, called sequential cutting, consists of cutting each unit that will

TABLE 14-12 Opening Formation Alternatives

Opening formation	Limitations	Advantages	Disadvantages
Sequential cutting	None	Low waste, ease of buck insertion	Time, some potential inaccuracy
In-place cutting	None	Cutting speed, accuracy	Potential waste, awkwardness of buck insertion, setting time, preplanning of rebar

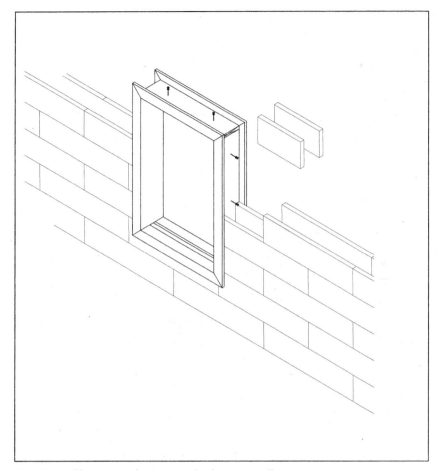

Figure 14-14 Units cut and set up to a buck sequentially.

fall along the opening's perimeter just before setting it. Figure 14-14 depicts the process. Waste tends to be low because each cutoff is immediately available for use in the next location where it will fit. Installing the buck is easy: one places it on the wall as soon as the forms reach sill level. Then one cuts each unit to precisely abut the buck, without otherwise altering the set sequence. Since the cuts are made in each unit separately, they may not align exactly around the opening. But in practice, careful work results in alignment that is more than adequate for most purposes.

The alternative is to build the form wall without an opening, then cut it out in place. Figure 14-15 shows this process. The cutting is fast, and it results in an especially accurate opening. It may produce greater waste, however. Time is also used to set unnecessary forms,

Figure 14-15 Cutting out a small opening in place. (*Portland Cement Association.*)

particularly when the opening is large. And setting the buck can be more difficult. Flanges must be left off one face of a protruding buck and then attached after it is in place. A recessed buck must be angled into the core of the formwork, or set into a wide-cut opening and covered with foam once in place. Setting horizontal reinforcing bars requires advance planning so that they come up to, but do not encroach on the intended opening.

In practice, in-place cutting is commonly used for two types of openings. The first is small ones. Small openings often require more precision, and pose only minor waste and logistic problems. The second type is irregular openings, such as round, diamond, or arched windows. These are difficult to cut accurately in a sequential fashion.

Formwork 291

TABLE 14-13 Irregular Buck Alternatives

Irregular buck	Limitations	Advantages	Disadvantages
Lumber	None	Fastening surface, rigidity	Time and labor, potential waste
Sheet material	None	Ease of formation	Lack of fastening surface, lack of rigidity

Irregular Bucks

Nonrectangular openings (Table 14-13) can be given a conventional lumber buck (flanged or stucco). Lumber bucks are generally strong, and serve as a fastening surface around the perimeter. However, buck construction can be time-consuming and generate a relatively large amount of waste, particularly if the opening includes curves or many short segments.

An alternative is to bend a thin sheet material into the desired shape. Figure 14-16 depicts this. Since such a buck will not be highly rigid, this is generally done in conjunction with in-place cutting of the opening (see preceding section). After the opening is cut, the buck is

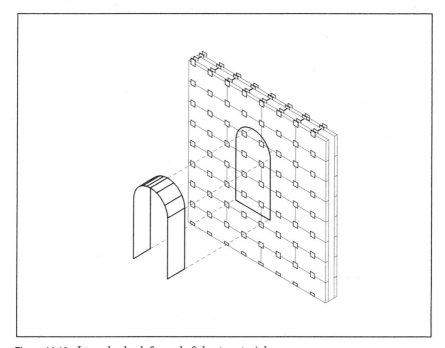

Figure 14-16 Irregular buck formed of sheet material.

glued to the formwork around its perimeter. The formwork cut from inside the opening can be replaced to hold the buck in shape. Since the buck is not usually wide enough to form a good fastening surface, workers must otherwise provide for attaching trim or other items around the opening.

Chapter 15
Concrete

Placing concrete into ICF formwork is much the same as when using conventional forms. But the differences are important to learn and follow.

Mix

Although good plans specify the compressive strength (psi) of the concrete to be used, other properties and the exact ingredients are generally left open. *One should always follow the recommendations of the design engineer or the ICF manufacturer in specifying concrete.* Most manufacturers offer precise rules for concrete specification based on experience with their product and on engineering analysis. The information provided in this section is intended only to assist specification decisions in areas where engineers' and manufacturers' instructions leave latitude.

Crews placing the concrete naturally tend to prefer mix specifications that result in concrete with a higher flow and lower cost. The major options are listed in Table 15-1.

Flow is an informal term referring to the ability of the concrete to move into and through small spaces. Concrete mixes of greater slump or smaller maximum aggregate sizes tend to flow better. This helps completely fill the small spaces present in some grid, post-and-beam, and narrow flat walls.

Standard residential mixes usually contain ¾-inch aggregate (maximum size). Their flow is often not sufficient to place them with a boom or line pump (see the following section), and may not be sufficient to penetrate small cavities easily. However, they tend to be the least expensive option, and they are familiar to all parties involved. They are often used in flat systems of at least 8-inch concrete thickness, placed with a chute or bucket.

TABLE 15-1 Concrete Mix Alternatives

Concrete mix	Limitations	Advantages	Disadvantages
Standard	Large-cavity systems, chute or bucket placement	Low cost, familiarity	Low flow
Pump mix	None	Increased flow	Cost
Higher proportion of fine aggregate (sand)	None	Increased flow	Cost
Extra water	None	Increased flow, low cost	Weakened concrete, reduced watertightness, increased pressure on forms, potential failure to meet specs
Plasticizer	None	Increased flow, strength, water tightness	Cost, unfamiliarity, possible quick hardening

Pump mix is an informal term for a concrete mix designed to flow through a concrete pump. Requesting it may lead to a slightly different combination of ingredients from supplier to supplier. It usually includes aggregates of up to ⅜ inches only, and often the proportions of the various ingredients are different from a standard mix. It has a higher flow, at a cost of about $1 to $2 per yard more than standard concrete.

One may also replace some of the larger aggregates with sand. For example, a mix of 60 percent stone aggregate and 40 percent sand might be reversed to proportions of 40 percent stone and 60 percent sand. Flow increases at a usually small cost.

Another option is to increase the slump of a given mix by adding extra water. *This is a potentially dangerous action, and should be undertaken only within ranges allowed by the building code, applicable standards, the project designers, and the ICF manufacturer.* The attraction is that water increases flow at no additional cost. Systems with narrow spaces in their core recommend slumps as high as 6 inches. Concrete suppliers told the desired psi and slump in advance can provide an appropriate mix. A big disadvantage of using water is that it raises the water-to-cement ratio. Beyond a minimum amount of water necessary to cause the concrete to cure, adding water (and thereby increasing the ratio) reduces the strength (psi) and watertightness of the final concrete. Adding water is a particularly dangerous step to take in the field, since it may reduce the performance of the concrete below what the designers have determined is necessary. In addition, more liquid concrete puts greater pressure on the forms.

A final alternative is to use a plasticizer, a substance designed to increase the slump of concrete without changing the water-to-cement ratio. A plasticizer does not have the detrimental effects of extra water. It adds some cost, and not all concrete vendors are familiar with it in detail. Familiarity is important, since most contractors must rely on the vendor to select the correct plasticizer and correct proportions. In addition, some plasticizers cause the concrete to harden suddenly after 1 to 3 hours. So quick placement can be essential. But for ICF projects its effects on the flow and quality of concrete are nearly ideal.

Placement Equipment

Four types of equipment are commonly used to place concrete into ICF formwork (Table 15-2). The least expensive is simply to use the chute provided with the concrete truck, as pictured in Figure 15-1. Since it works by gravity, this method is limited to pours that are below grade. Control of the concrete flow is only moderately precise. To prevent the material from flowing so quickly that it might damage the formwork, some crews assign one worker permanently to holding a shovel at the end of the chute in order to slow the concrete before it drops. This method of placement also suffers from occasional interruptions in work as the truck and the chute are repositioned. Access around the perimeter is necessary to deposit all concrete within 5 feet of its final location. The truck will need to make multiple trips around

TABLE 15-2 Placement Equipment Alternatives

Placement equipment	Limitations	Advantages	Disadvantages
Chute	Below grade	Low cost	Moderate control, intermittent operation, risks of equipment near excavation
Boom pump	None	Continuous operation, good positioning, physical ease	Difficulty of control, high cost, need for special mix, waste, potential for formwork damage
Line pump	None	Good control, continuous operation	Weight of line, moderate cost, need for special mix, modest waste, potential for formwork damage
Bucket	None	Very good control, good positioning	Intermittent operation, moderate cost

296   Assembly

Figure 15-1   Placing concrete with a standard chute.

because most pours consist of 2 or 3 lifts of 2 to 6 feet. And chute placement carries the risks of placing heavy equipment near the edge of an excavation. But the savings motivate most crews to use it in nearly all situations where it is logistically feasible.

Boom pumps are common in above-grade placement. They normally consist of a concrete pump and a boom mounted on a truck trailer, as in Figure 15-2. A hose runs the length of the boom and beyond, the end dangling back down toward the ground. The pump forces concrete into the hose, where it climbs the boom and falls down the free end. Reducers (fittings that reduce the diameter of the hose) and angle fittings at the end of the hose are necessary to slow the concrete flow. The position of the boom is hydraulically and continuously adjustable by an operator below.

The end of the hose of a boom pump can be positioned fairly precisely, and work must rarely wait on moving it. All of the heavy work is done by machine. However, there is a lag of a couple of seconds between the time the crew on the scaffolding signals directions to the operator and the time the pump responds. This requires some tricky compensation by the crew. The intermittent action of the pump also causes the line to surge, requiring steadying. If the hose is put in contact with the formwork, the surging might cause damage. The boom pump is also usually the most expensive option (around $500 per half-day in many places). It requires a pump mix of concrete. This

Figure 15-2 Placing concrete with a boom pump. (*Reddi-Form Inc.*)

must be specified upon ordering and usually adds about $1 to $2 per yard to the concrete cost. Using the pump can also waste some concrete: up to 1 yard is necessary to prime the pump, and that must be used elsewhere or dropped on the ground after placement is done.

A somewhat less expensive option above grade is the line pump pictured in Figure 15-3. It is ordinarily pulled as a trailer on its own wheels. The hose (or "line") lies on the ground. One or two workers hold its end up over the formwork to drop concrete into the core. A simple switch to turn it on or off is usually connected to the pump by a long wire, so that control is possible from the end of the hose. Control over the position of the line is good, and the lag between the crew's desire to turn off concrete and the result is shorter than with a boom pump. But the line is heavy, requiring full effort from at least one crew member to hold and control it. If it is set on the formwork carelessly it might damage it. Line pumps also require a pump mix. They are usually somewhat slower than boom pumps and cost a little less.

A concrete bucket (Figure 15-4) suspended from a general-purpose crane is a third option used above grade. The bucket is heavy steel, large enough to hold about a cubic yard at a time. At the bottom is

298 Assembly

Figure 15-3 Line pump. (*Independent Concrete Pumping.*)

Figure 15-4 Placing concrete for a floor with a bucket. (*Portland Cement Association.*)

usually a hand-controlled flap that is opened or closed to start or stop the flow of concrete. The bucket rests on the ground while a concrete truck fills it. The crane operator hoists the bucket over the wall as directed by the crew. Positioning is precise, and the flap allows the worker pouring the concrete to start and stop flow at will. Buckets can handle any common concrete mix. The costs of a bucket and crane are usually about the same as those of a line pump. Operation must be interrupted periodically to refill the bucket, and some contractors report that this makes the pour slower than with a chute or boom pump.

Consolidation

While placing the concrete it is important to consolidate it. All methods of consolidation involve vibrating the concrete or the formwork, so one must be careful not to skew the alignment of the wall or damage the forms.

Table 15-3 lists popular methods of consolidation. A common one is to tap the wall in the sections containing concrete as they are filled. Often this is done by placing a lumber scrap against the formwork directly over a tie and tapping with a hammer. This procedure is simple and uses available tools. It is easy for the workers to miss spots if they are not conscientious, however.

One alternative is to plunge a steel reinforcing bar or long piece of lumber up and down a few times in the concrete in each vertical cavity immediately after placing concrete into it. Such "rodding" has the same basic advantages and disadvantages as tapping with a hammer. In addition, rodding carelessly may damage the formwork.

TABLE 15-3 Consolidation Alternatives

Consolidation	Limitations	Advantages	Disadvantages
Tapping	None	Simplicity	Potential inconsistency
Rodding	None	Simplicity	Potential inconsistency, potential for form damage
Form vibration	Approved brands	Potential thoroughness	Need for appropriate tool, potential for form damage
Concrete vibration	Approved brands	Thoroughness	Need for special equipment, equipment cost, potential for form damage
None	Large-cavity brands	Speed, ease	Need for special mix, potential for voids

Figure 15-5 Consolidating concrete with a vibrator. (*Lite-Form Inc.*)

Some crews vibrate the formwork with a convenient tool placed against the wall over a tie. A reciprocating saw with the blade removed is a common one. This may be more consistent and thorough than pure hand work. But it has the potential of damaging the formwork or causing blowouts. Some manufacturers specifically disallow it with their systems.

Concrete vibrators are electric devices with a long rod (or "needle") that is inserted into the concrete. Figure 15-5 pictures one. This method of consolidation can be very thorough. However, it requires a special, potentially expensive tool. Improperly used it can also damage or blow out at least some types of formwork. Some manufacturers disallow the use of vibrators or restrict their sizes.

Users of very large-cavity systems may not bother to consolidate. The advantage is time and cost savings. However, this runs the risk of voids in the final wall. The risk is least when a high-flow mix (see "Mix" earlier in this chapter) is placed into formwork with wide, flat cavities.

Sloped Pours

Some designs require a slope on the concrete, usually to build ICF gable or gambrel end walls. One method of doing this is directly (see Table 15-4 and Figure 15-6). The crew places concrete into formwork cut at the desired, sloped roofline, and works it even with the top of the form at all points, using trowels. As it sets they insert anchor bolts or metal straps, as called for by the roof design. Of course this method

Concrete 301

TABLE 15-4 Angled Pour Alternatives

Angled pour	Limitations	Advantages	Disadvantages
Direct	Modest inclines and low slumps	Simplicity, low advance preparation, ease of concrete placement	Potential variation, difficulty of evening concrete
Successive blocks	None	Precision, little work during pour	Advance preparation, some extra consolidation
Successive plates	None	Precision, ease of concrete placement	Work required during pour, advance preparation
Fixed blocks	Moderate inclines and slumps	Modest advance preparation	Potential variation

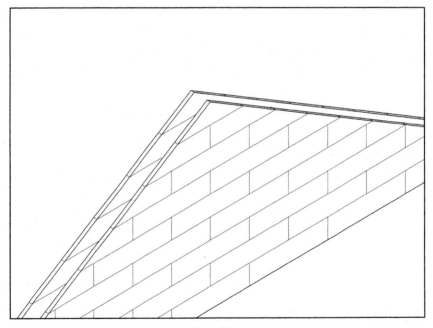

Figure 15-6 Gable end formwork ready for a direct pour.

302 Assembly

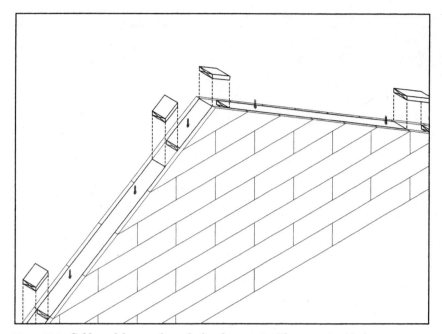

Figure 15-7 Gable end formwork ready for placement with successive blocks.

is only feasible with modest slopes and lower-slump concretes. It is simple and requires little preparation. However, it relies on the accuracy of the crew's handwork. It may also require the workers to stay at the job for a while to push the concrete back into position repeatedly, until it sets. Trying to work around the embedded fasteners and hold them in position can, unfortunately, impede this process.

A popular alternative is the successive blocks method. It is depicted in Figure 15-7. Under it, one preinstalls top plates in the wall, leaving gaps of a few inches between each one. The plates are generally glued into place or (for recessed plates) fastened through the foam with insulation nails or insulation screws. They hold preinstalled anchor bolts for a secure concrete connection. The crew fills the wall through the gaps. Some extra consolidation may be necessary to move the concrete adequately sideways from these widely spaced points. Once the concrete reaches the lowest of the gaps, the crew plugs them with wooden blocks fastened to the plates or formwork and continues filling from the remaining gaps. When the wall is full, the last gap (at the peak) is plugged to create a continuous line of top plates.

Successive blocks produce a precise top of wall with a moderate level of effort during the pour. But plates and blocks must be prepared in advance.

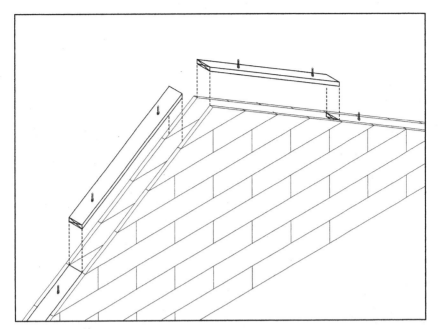

Figure 15-8 Successive plates for use in filling a gable end.

A related alternative is the method of successive plates (Figure 15-8). Under this procedure the crew attaches only the lowest plates to the formwork in advance. The rest are precut and ready nearby. All plates have preinstalled anchor bolts. The crew fills through the uncovered, higher portion of the wall. Once the level of the concrete approaches the top of the plates on the wall, the next higher plates go into position and filling continues. The last plates are sized to leave a small opening at top for the last stage of filling. A block fills this in after the wall is full.

Pluses and minuses of this method are similar to those of successive blocks. But compared to that method, concrete is a little easier to place, yet more work is involved during the pour.

The last alternative depends on small blocks recessed flush with the top of the formwork every few feet. Figure 15-9 depicts this "fixed blocks" arrangement. They hold preinstalled anchor bolts or metal straps to fasten the top plates later to the top of the wall. Concrete goes between the blocks and crew members even it out with trowels, as in the direct method. The advantage of using the blocks is that they help hold the angle on the concrete and hold the anchor bolts in position without impeding the trowel work. Otherwise this method is similar to the direct pour.

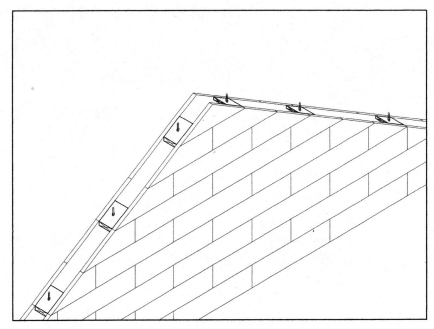

Figure 15-9 Gable end formwork ready for concrete placement with fixed blocks.

Chapter 16

Attachments

The contractor has important alternatives for making some of the attachments to an ICF wall.

Floor Joist Pocketing

When building a pocketed floor deck (discussed in Chapter 7 under "Floor Decks"), there are two methods in use for inserting the ends of the floor joists into the concrete wall. Table 16-1 compares them.

A common method is to form the pockets with small wooden blocks cut for the purpose (Figure 16-1). Workers install the blocks and metal anchors in the formwork at the measured joist locations before the pour. After the concrete hardens they knock out the blocks and set a joist into each of the pockets formed, then nail the anchor straps to their joists.

Direct joist insertion is an alternative that works well with ICFs. Workers cut precise slots in the formwork to hold the end of each joist. When steel members are used (as depicted in Figure 16-2), the slot is a C shape to match the member's profile. The crew then slides the joist ends into the slots and attaches the required metal anchors. The crew braces the joists temporarily with a horizontal support just below the

TABLE 16-1 Joist Pocketing Alternatives

Pocketing	Limitations	Advantages	Disadvantages
Pocket blocks	None	Familiarity	Materials required
Direct insertion	None	Low materials, available tool	Unfamiliarity, mess, physical effort, imprecision

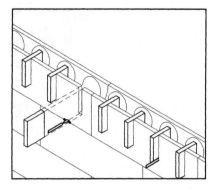 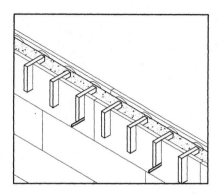

Insert blocks and t-straps   Place concrete

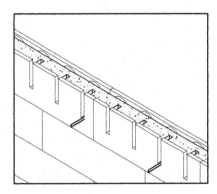 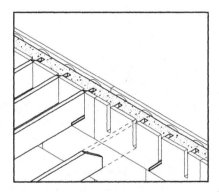

Remove blocks   Insert joists and fasten straps

Figure 16-1 Constructing a pocketed floor with pocket blocks.

slots (such as a stud attached to the surface of the formwork) or with vertical studs set under the joists that extend down to the ground. This arrangement usually supports the joists well enough for the crew to complete the deck and stand on top of it for the pour.

Direct insertion uses less materials and labor. It also allows for building the entire floor deck early and using it in lieu of scaffolding during the pour. This may allow the adoption of simpler, lower scaffolding. The major disadvantage is the unfamiliarity of the technique and limited experience with it.

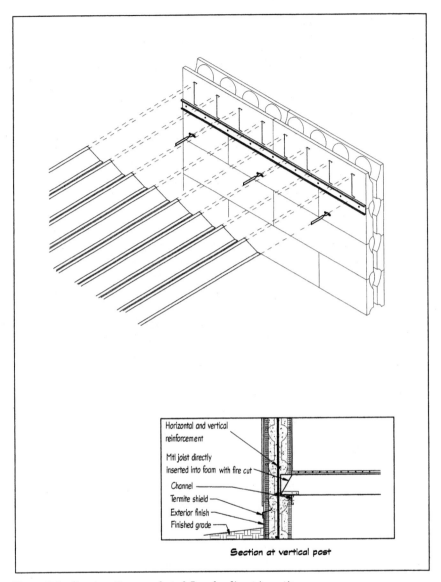

Figure 16-2 Constructing a pocketed floor by direct insertion.

TABLE 16-2 Surface Cutting Alternatives

Surface cutting	Limitations	Advantages	Disadvantages
Router	None	Speed, precision, familiarity	Equipment cost, need for special tool, potential tool damage, mess, weight
Circular saw	None	Speed, familiarity	Weight, mess
Cold knife	Pure foam	Low equipment cost, familiarity, available tool	Slowness, mess, physical effort, imprecision
Hot knife	Pure foam	Moderate speed, precision, potential to replace foam	Equipment cost, need for special tool

Surface Cutting

Placing electrical and plumbing services along an ICF wall most often begins with cutting channels and rectangles in the foam surface for the lines and boxes. A variety of tools is available to do the job (Table 16-2).

All methods either cut the foam mechanically (as wood is cut) or cut with heat. All will work in pure foams, but only heavy mechanical cutting will penetrate the foam-cement composites effectively.

An ordinary router makes fast, precise cuts and is a familiar tool. However, it is tiring to hold over extended periods because of its weight, it is a moderately expensive tool, and most electrical and plumbing crews will have to add it to the supplies they bring to the project. It usually requires a special long bit to make a sufficiently deep cut. Hitting metal (such as a steel fastening surface) can damage the bit. Some routers throw off long, unusable streamers of the foam.

It is possible to cut channels in two passes with an ordinary circular saw. Holding it at separate angles each time creates a V channel. This is fairly fast and uses a familiar tool that is already on site. However, holding the saw up can be tiring. There is also a moderate mess.

An ordinary construction knife is cheap and familiar. Most crews carry one as a matter of course. But it is also slow, particularly messy, tiring because of the effort required, and tends to make uneven cuts.

A few companies sell electric hot knives designed to cut foam. One is pictured in Figure 16-3. When plugged in, the knives' precisely shaped and sized blades cut fairly quickly. They are so exact that the channel cutouts remain in tact as long, thin pieces that can be replaced later. (It may be desirable to put the cutouts back in the channels after cable is inserted to maintain full insulation.) However, the tool is expensive (a few hundred dollars) and must be specially ordered and kept for this single purpose.

Figure 16-3 Channel containing electrical cable (center) cut with a hot knife (lower right). (*Lite-Form Inc.*)

Appendix A

Metric Conversion Factors

The following list provides the conversion relationships between U.S. customary units and SI (International System) units. The proper conversion procedure is to multiply the specified value on the left (primarily U.S. customary values) by the conversion factor exactly as given and then round to the appropriate number of significant digits desired. For example, to convert 11.4 feet to meters: $11.4 \times 0.3048 = 3.47472$, which rounds to 3.47 meters. Do not round either value before performing the multiplication, as accuracy would be reduced. A complete guide to the SI system and its use can be found in ASTM E 380, *Metric Practice*.

To convert from	to	multiply by	
Length			
inch (in)	micrometer (μ)	25,400	E*
inch (in)	centimeter (cm)	2.54	E
inch (in)	meter (m)	0.0254	E
foot (ft)	meter (m)	0.3048	E
yard (yd)	meter (m)	0.9144	
Area			
square foot (sq ft)	square meter (sq m)	0.09290304	E
square inch (sq in)	square centimeter (sq cm)	6.452	E
square inch (sq in)	square meter (sq m)	0.00064516	E
square yard (sq yd)	square meter (sq m)	0.8361274	
Volume			
cubic inch (cu in)	cubic centimeter (cu cm)	16.387064	
cubic inch (cu in)	cubic meter (cu m)	0.00001639	
cubic foot (cu ft)	cubic meter (cu m)	0.02831685	
cubic yard (cu yd)	cubic meter (cu m)	0.7645549	

To convert from	to	multiply by
gallon (gal) Can. liquid	liter (L)	4.546
gallon (gal) Can. liquid	cubic meter (cu m)	0.004546
gallon (gal) US liquid†	liter (L)	3.7854118
gallon (gal) US liquid	cubic meter (cu m)	0.00378541
fluid ounce (fl oz)	milliliter (mL)	29.57353
fluid ounce (fl oz)	cubic meter (cu m)	0.00002957

Force

kip (1000 lb)	kilogram (kg)	453.6
kip (1000 lb)	newton (N)	4448.222
pound (lb)	kilogram (kg)	0.4535924
pound (lb)	newton (N)	4.448222

Pressure or stress

kip per square inch (ksi)	megapascal (MPa)	6.894757
kip per square inch (ksi)	kilogram per square centimeter (kg/sq cm)	70.31
pound per square foot (psf)	kilogram per square meter (kg/sq m)	4.8824
pound per square foot (psf)	pascal (Pa)‡	
pound per square inch (psi)	kilogram per square centimeter (kg/sq cm)	0.07031
pound per square inch (psi)	pascal (Pa)‡	6894.757
pound per square inch (psi)	megapascal (MPa)	0.00689476

Mass (weight)

pound (lb)	kilogram (kg)	0.4535924
ton (2000 lb)	kilogram (kg)	907.1848
grain	kilogram (kg)	0.0000648

Mass (weight) per length

kip per lineal foot (klf)	kilogram per meter (kg/m)	0.001488
pound per lineal foot (plf)	kilogram per meter (kg/m)	1.488

Mass per volume (density)

pound per cubic foot (pcf)	kilogram per cubic meter (kg/cu m)	16.01846
pound per cubic yard (lb/cu yd)	kilogram per cubic meter (kg/cu m)	0.5933

Temperature

degree Fahrenheit (°F)	degree Celsius (°C)	$t_c = (t_f - 32)/1.8$
degree Fahrenheit (°F)	kelvin (K)	$t_k = (t_f + 459.7)/1.8$
Kelvin (K)	degree Celsius (°C)	$t_c = t_k - 273.15$

Energy and heat

British thermal unit (Btu)	joule (J)	1055.056	
calorie (cal)	joule (J)	4.1868	E

To convert from	to	multiply by	
Btu/(°F · h · sq ft)	W/(m² · K)	5.678263	
kilowatt-hour (kWh)	joule (J)	3,600,000	E
British thermal unit per pound (Btu/lb)	calories per gram (cal/g)	0.55556	
British thermal unit per hour (Btu/h)	watt (W)	0.2930711	

Power

horsepower (hp) (550 ft · lb/s)	watt (W)	745.6999	E

Velocity

mile per hour (mph)	kilometer per hour (km/h)	1.60934
mile per hour (mph)	meter per second (m/s)	0.44704

Permeability

darcy	centimeter per second (cm/s)	0.000968
feet per day (ft/day)	centimeter per second (cm/s)	0.000352

*E indicates that the factor given is exact.
†One U.S. gallon equals 0.8327 Canadian gallon.
‡A pascal equals 1000 newtons per square meter.

NOTES:

One U.S. gallon of water weighs 8.34 pounds (U.S.) at 60°F.
One cubic foot of water weighs 62.4 pounds (U.S.).
One milliliter of water has a mass of 1 gram and a volume of one cubic centimeter.
One U.S. bag of cement weighs 94 lb.

The prefixes and symbols listed below are commonly used to form names and symbols of the decimal multiples and submultiples of the SI units.

Multiplication factor	Prefix	Symbol
$1,000,000,000 = 10^8$	giga	G
$1,000,000 = 10^5$	mega	M
$1000 = 10^3$	kilo	k
$1 = 1$	—	—
$0.01 = 10^{-2}$	centi	c
$0.001 = 10^{-3}$	milli	m
$0.000001 = 10^{-6}$	micro	μ
$0.000000001 = 10^{-9}$	nano	n

Appendix B

Canadian Equivalents of Referenced U.S. Standards

U.S. Standard	Title	Canadian Standard
ACI 318	Building Code Requirements for Structural Concrete	CSA A23.3-94
ASTM C578	Standard Specification for Rigid, Cellular Polystyrene Thermal Insulation	CAN/CGSB 51.20-M87
ASTM C236	Standard Method for Steady-State Thermal Performance of Building Assemblies by Means of a Guarded Hot Box	ASTM C236*
ASTM E119	Standard Test Method for Fire Tests of Building Construction Materials	CAN/ULC-S101-M89
MEC 502.1.2	Provisions for Thermal Mass, Section 502—Building Envelope Requirements, Model Energy Code	None
PFCD/GS4-8/95	Spray Polyurethane Foam for Building Envelope Insulation and Air Seal	CAN/CGSB 51.20-M87

*This ASTM standard is also referenced in Canada.

Appendix C

Directory of Product and Information Sources

General information

Canadian Portland Cement Association
116 Albert Street, Suite 609
Ottawa, ON K1P 5G3, Canada
Phone: (613) 236-9471
Fax: (613) 563-4498

Insulating Concrete Form Association
960 Harlem Ave. #1128
Glenview, IL 60025
Phone: (708) 657-9730
Fax: (708) 657-9728

National Ready Mixed Concrete Association
Promotion Department
900 Spring Street
Silver Spring, MD 20190
Phone: (301) 587-1400
Fax: (301) 585-4219

Portland Cement Association
5420 Old Orchard Road
Skokie, IL 60077-1083
Phone: (708) 966-6200
Fax: (708) 966-8389

ICF Systems Available in the United States

Amhome
Amhome U.S.A. Inc.
P.O. Box 1492
Land O'Lakes, FL 34639
Phone: (813) 996-4660
Fax: (813) 996-5452

Blue Maxx
AAB Building Systems Inc.
840 Division Street
Cobourg, ON K9A 4J9, Canada
Phone: (905) 373-0004
(800) 293-3210
Fax: (905) 373-0002

Diamond Snap-Form
AFM Corporation
P.O. Box 246

Excelsior, MN 55331
Phone: (800) 255-0176
Fax: (612) 474-2074

ENER-GRID
ENER-GRID Building
  Systems Inc.
6847 S. Rainbow Road
Buckeye, AZ 85326
Phone: (602) 386-2232
Fax: (602) 386-3298

ENERGY LOCK
Energy Lock Inc.
521 West 3560 South
Salt Lake City, UT 84115
Phone: (801) 288-1199
Fax: (801) 288-1196

Featherlite
Featherlite Inc.
18 Turtle Creek Drive
Tequesta, FL 33469
Phone: (407) 575-1193
Fax: (407) 575-1959

Fold-Form
See Lite-Form.

GREENBLOCK
GREENBLOCK
  WorldWide Corp.
P.O. Box 749
Woodland Park, CO 80866
Phone: (719) 687-0645
Fax: (719) 687-7820

Insulform
Western Insulfoam
Division of Premier
  Industries, Inc.
19041 80th Avenue South
Kent, WA 98032
Phone: (206) 242-9424
Fax: (206) 251-8405

I.C.E. Block
North American I.C.E. Block
  Association
4427 S. Polaris Ave.

Las Vegas, NV 89103
Phone: (800) ICE-BLKS

KEEVA
KEEVA International Inc.
1854 North Acacia Street
Mesa, AZ 85213
Phone: (602) 827-9894
Fax: (602) 827-9697

Modu-Lock
See Quad-Lock.

Lite-Form
Lite-Form Inc.
P.O. Box 774
1210 Steuben Street
Sioux City, IA 51102
Phone: (712) 252-3704
Fax: (712) 252-3259

Polycrete
Polycrete Industries Inc.
435 rue Trans-Canada
Longueuil, QUE J4G 2P9,
  Canada
Phone: (514) 646-3825
Fax: (514) 646-4880

Polysteel
American Polysteel Forms
5150-F Edith Avenue
Albuquerque, NM 87107
Phone: (800) 977-3676
Fax: (505) 345-8154

Quad-Lock
Quad-Lock Building
  Systems
3873 Airport Way,
  Suite 525
Bellingham, WA 98226
Phone: (360) 671-3911
Fax: (360) 671-7639

RASTRA
RASTRA/Environmental
  Building Technologies
100 South Sunrise Way,
  Suite 289

Palm Springs, CA 92262
Phone: (619) 778-6593
Fax: (619) 778-4917

Reddi-Form
Reddi-Form Inc.
593 Ramapo Valley Road
Oakland, NJ 07436
Phone: (201) 405-2030
Fax: (201) 405-1987

REWARD
3-10 Insulated Forms L.P.
P.O. Box 46790
Omaha, NE 68128
Phone: (402) 592-7077

R-FORMS
R-FORMS/Owens-Corning
3 Century Drive
Parsippany, NJ 07054
Phone: (201) 267-1605
Fax: (201) 267-1495

SmartBlock VWF and SF 10
American ConForm
Industries Inc.
1820 South Santa Fe Street
Santa Ana, CA 92705
Phone: (800) CONFORM
Fax: (714) 662-0405

ThermoFormed
ThermoFormed Block Corp.
P.O. Box 2454
Fort Collins, CO 80522-2454
Phone: (800) 821-0855

THERM-O-WALL
New England Foam Form
26 North Raymond Road
Gray, ME 04039
Phone: (207) 657-2222

Warm Walls of Minnesota
425 Second Avenue, S.W.
Oronoco, MN 55960
Phone: (800) 424-9255

ICF Systems Available in Canada Only

CONSULWAL
CONSULWAL
2668 Mount Albert Road East,
  RR2
Queensville, ON L0G 1R0,
  Canada
Phone: (905) 853-2027
Fax: (905) 478-2618

ENVI-R-FORM
ENVI-R-FORM Building
  Systems Inc.
Unit Z, 2700 Barnet Highway
Coquitlan, BC V3B1B8, Canada
Phone: (604) 942-5488
       (800) 701-2223
Fax: (604) 936-7908

I-Form
International Form Building
  Systems
11711 #5 Road Unit #8
Richmond, BC V7A 4E8,
  Canada
Phone: (604) 448-8601
Fax: (604) 448-8647

In-form
In-form Canada
1430-1188 West Georgia Street
Vancouver, BC V6E 4A2,
  Canada
Phone: (604) 682-6200

KEPSYSTEM
KEPSYSTEM Inc.
140, rue St.-Eustache, Room 301
St.-Eustache, QUE J7R 2K9,
  Canada
Phone: (514) 472-3560
Fax: (514) 472-3685
Fax: (604) 682-7388

Index

ACH (*see* air changes per hour)
ACI (*see* American Concrete Institute)
adhesives, 38, 191, 275
air changes per hour, 7, 8, 243, 250
air infiltration, 7, 8–9, 10, 241, 247, 249–250, 256
air quality, 88, 179, 181, 192, 243
American Concrete Institute, 16, 203, 253
American Society for Testing and Materials, 68–77
American Society of Heating, Refrigeration, and Air Conditioning Engineers, 242
Amhome, 50, 55–57, 98, 106, 112, 187
anchor bolts (*see* fasteners)
ASHRAE (*see* American Society of Heating, Refrigeration, and Air Conditioning Engineers)
ASTM (*see* American Society for Testing and Materials)

beam, 32, 40, 44
beam cavity (*see* cavities)
bending moment, 209–212, 225–231, 232
block systems, 29, 31, 33, 277, 282
  (*See also* Blue Maxx; ENERGY LOCK; flat block systems; Featherlite; Fold-Form; GREENBLOCK; grid block systems; Insulform; KEEVA; Modu-Lock; Polysteel; Reddi-Form; REWARD; SmartBlock SF 10; SmartBlock VWF; Therm-O-Wall)
blower door test, 8
Blue Maxx, 51, 59, 60, 98, 112, 188
BOCA (*see* Building Officials Conference of America)
bond-beam block (*see* lintel unit)

boom pump (*see* concrete, placement)
bracing:
  of floors, 160, 161
  of formwork, 47, 264, 269, 283–284
  of walls, 232
  (*See also* bracing and scaffolding systems, top bracing)
bracing and scaffolding systems, 47, 48, 49–53, 266, 267, 283–284
brick ledge unit, 46, 47, 186
bucket (*see* concrete, placement)
bucks, 135, 139–146, 190, 264, 265, 269, 289–291
  channel, 145–146, 190
  flanged (*see* protruding)
  protruding, 142, 144
  recessed, 139–143
  stucco (*see* recessed)
building codes, 106–108, 119, 203, 253
  R-value provisions, 11, 248
  insect provisions, 93, 134
  seismic provisions, 235
Building Officials Conference of America, 98–99, 107, 203, 235

CABO (*see* Council of American Building Officials)
Canadian Construction Materials Centre, 98–99, 107
Canadian Standards Association, 16, 204
cavities, 38, 46, 61, 217
  alignment, 38, 262, 264, 274
  categories, 38, 39
  positioning, 155, 272
CCMC (*see* Canadian Construction Materials Centre)

concrete:
  compressive strength, 205–206, 215, 253–254, 255, 256, 293
  cross section thickness, 111–113, 187–188, 205, 207, 208, 215, 219, 221, 232
  flow, 293–295
  placement, 269, 293–304
  shear strength, 217, 256
  slump, 293
  usage, 187–188, 189
concrete pumps, 189
Construction Technology Laboratories, 246, 255
core, 39
  (See also cavities)
corner-corner setting, 277–278
corner-in setting, 277–278
corners, 11–12, 40, 46, 49–53, 59, 61, 62, 105–106
  creation of, 41–42, 43, 288
  hinged, 41, 288
    (See also hinge unit)
  precut, 41, 186, 288
  preformed, 41–42, 43, 57, 186, 288
  user-cut, 41–42, 43, 288
cost, 97, 177–193, 199
  of concrete, 113, 293–294, 297
  of construction, 3–6
  crew experience and, 5–6, 178, 181, 191–192
  of exterior finish, 5, 178–179
  of HVAC equipment, 3, 5, 6, 179, 181, 192
  operating, 7
  of spans and projections, 13–14
  trends, 6
  of units, 101, 113, 178–179
  of wall, 128
  (See also specific item, as, foam, cost)
Council of American Building Officials, 248
crews, 271–272
  learning by, 15, 154, 277
  (See also cost, crew experience and)
CTL (see Construction Technology Laboratories)
curved walls, 12, 180, 273, 284–287
cylinders (see cavities)

Diamond Snap-Form, 50, 57, 98, 112, 187
dimensions of units, 30, 98–99, 102, 142, 187–189, 197

dimensions of units (Cont.):
  and building dimensions, 109–111
  consistency, 100
direct joist insertion, 305
double running bond, 280–281
dry set, 275–276
durability, 20–22
  of concrete, 20
  finish and, 21
  of foam, 20–21

earthquake resistance, 17, 102, 199, 200
  provisions, 179, 180
eccentricity, 208, 232
electrical wiring (see utility lines)
emissions, 23–24
  in fire, 16
end pieces (see stops)
ENER-GRID, 49, 55, 98, 112, 187
ENERGY LOCK, 53, 62–63, 99, 113, 188
EPS (see foam, expanded polystyrene)
expanded polystyrene (see foam, expanded polystyrene)
extruded polystyrene (see foam, extruded polystyrene)

face shells, 37, 59, 60, 106, 113
faces, 37, 178
fasteners, 191
  for floor decks, 154–159
  for interior walls, 166–167
  for roofs, 163–165, 190–191
  for top plate, 125, 300–303
  (See also fastening surfaces, fasteners for)
fastening surfaces, 49–53, 56, 59, 61, 62, 98–99, 103–104, 262
  defined, 38
  fasteners for, 38, 39
Featherlite, 53, 62–63, 99, 113, 188
finishes, 242, 244
  exterior, 46, 89–91, 116, 117, 171–176, 178–180, 270
  interior, 164, 171–172, 270
  (See also specific finishes, as stucco)
fire cut, 155–159
fire resistance:
  of foam, 16, 67, 69, 75–77, 78, 79–80
  provisions for, 85, 117

fire resistance (*Cont.*):
  of walls, 16, 84–85, 102–103
  (*See also* flame spread index; smoke developed index, Steiner Tunnel test)
fire wall test, 16, 84
fixed blocks, 303
flame spread index, 16, 69, 77, 78
flat block systems, 35, 46, 51–52, 59–60
  (*See also* Blue Maxx; Fold-Form; GREENBLOCK; SmartBlock VWF)
flat panel systems, 33, 41, 48–49, 282
  [*See also* Lite-Form (preassembled); R-Forms]
flat plank systems, 35, 41, 48, 50–51, 57–59, 283, 285
  [*See also* Diamond Snap-Form; Lite-Form (unassembled); Polycrete; Quad-Lock]
flat systems, 31, 33, 39, 101, 109, 117, 120, 204, 207, 212, 219–221, 222, 293
  [*See also* Blue Maxx; Diamond Snap-Form; flat block systems; flat panel systems; flat plank systems; Fold-Form; GREENBLOCK; Lite-Form (preassembled); Lite-Form (unassembled); Polycrete; Quad-Lock; R-Forms; SmartBlock VWF]
floors, 153–163, 199, 268
  construction, 284, 305–307
  grid systems and, 155
  post-and-beam systems and, 155
Fold-Form, 51, 59, 60, 98, 106, 112, 188
foam, 49–53, 104–105
  categories, 65
  composite, 49–50, 55, 68, 79–80, 98, 103, 190
  cost, 67–69, 69, 78–79
  density, 66–71, 79
  expanded polystyrene, 50–53, 67, 68–78, 98–99
  extruded polystyrene, 67, 68–78
  polyurethane, 53, 62, 67–68, 99
  specifications, 66, 68–71, 78
  strength, 69, 72–74, 78, 79
  toxicity, 77, 79–80
  (*See also* specific properties, such as: fire resistance; R-value)
foam gap (*see* insects, provisions for)
forces (*see* loads)
formwork, 32–40
  damage, 256, 269, 296, 299–300
  setting, 261–264, 273–281

frames:
  nonsway, 210, 212
  sway, 210
furring strips, 168–169, 171–172, 173–176, 178–180

GREENBLOCK, 51, 59, 60, 98, 113, 188
grid block systems, 36, 52–53, 61–62
  (*See also* I.C.E. Block; Insulform; Modu-Lock; Polysteel; Reddi-Form; REWARD; SmartBlock SF 10; Therm-O-Wall)
grid panel systems, 34, 49–50, 55–56
  (*See also* ENER-GRID; RASTRA)
grid systems, 31–32, 33, 39, 101, 109, 111, 117, 121, 133, 204, 207, 212, 217, 219–221, 222, 225, 262, 293
  interrupted, 31–32
  uninterrupted, 31, 40
  screen (*see* grid systems, interrupted)
  waffle (*see* grid systems, uninterrupted)
  (*See also* ENER-GRID; grid block systems; grid panel systems; grid plank systems; I.C.E. Block; Insulform; Modu-Lock; Polysteel; RASTRA; Reddi-Form; REWARD; SmartBlock SF 10; Therm-O-Wall)
guarded hot box test, 8, 242
guides, 276

Hambro floor system, 159, 160
heat capacity, 243–245
heating, ventilating, and air conditioning (HVAC):
  load, 244–247
  sizing, 6, 179, 181, 241, 249–250, 256
  (*See also* cost, of HVAC equipment)
hinge unit, 12, 42, 44
hole plug, 46, 62
horizontal cavities (*see* cavities)
horizontal member (*see* beam)
hot knife, 308
hurricane straps (*see* fasteners)
HVAC (*see* heating, ventilating, and air conditioning)

ICBO (*see* International Conference of Building Officials)
I.C.E. Block, 52, 61, 99, 113, 188

Index

insects:
  resistance to, 18–19, 91–94, 102–103
  provisions for, 18–19, 92–94, 128–134, 142, 266
inspection gap (*see* insects, provisions for)
Insulform, 52, 61, 62, 99, 106, 113
interconnect, 30, 31, 37–38, 49–53, 54, 55, 57, 59, 61, 62, 281–283, 286
International Conference of Building Officials, 98–99, 107, 203, 235

KEEVA, 53, 62–63, 99, 113, 188

lifting equipment, 190
line pump (*see* concrete, placement)
lintel, 44, 197
  structural design, 198, 206, 217–221, 233–234
lintel block (*see* lintel unit)
lintel unit, 44–46, 62, 219
Lite-Form (nonassembled), 50, 57, 98, 112, 187
Lite-Form (preassembled), 48, 49, 54, 106, 112, 188
loads, 204–205, 231, 233–236
  axial, 207–212, 225–231, 232, 234, 237–239
  shear, 214–218, 219–225, 232–233

maintenance, 20
MEC (*see* Model Energy Code)
Model Energy Code, 248
Modu-Lock, 52, 61–62, 99, 113, 188
moisture resistance:
  of foam, 69, 70, 72, 74–76, 102–103
  provisions for, 17–18, 91, 117, 142, 155, 159, 175
  of walls, 17, 87–91, 119
  (*See also* windows, and moisture)
moment (*see* bending moment)
moment magnifier, 210–214

nonsway frames (*see* frames, nonsway)
North Carolina State Building Code (*see* building codes, insect provisions)

Occupational Safety and Health Administration, 23, 47, 266, 284

openings, 11, 15, 47, 243
  formation, 12–13, 135–146, 264, 288–292
  irregular, 13
  positioning, 110–111
OSHA (*see* Occupational Safety and Health Administration)

panel systems, 29, 30, 33, 273
  [*See also* Amhome; flat panel systems; grid panel systems; Lite-Form (preassembled); post-and-beam panel systems; R-Forms ENER-GRID; RASTRA; ThermoFormed]
PCA (*see* Portland Cement Association)
perimeter setting, 277, 279, 288
plank systems, 29, 30, 33, 42, 54, 273, 286
  [*See also* Diamond Snap-Form; flat plank systems; Lite-Form (unassembled); Polycrete; Quad-Lock]
plasticizer, 295
plumbing (*see* utility lines)
Polycrete, 51, 57–59, 98, 112, 187–188
Polysteel, 52, 61, 99, 113, 188
Portland Cement Association, 255
post, 32, 40
post cavity (*see* cavities)
post-and-beam block systems, 36, 40, 53, 62–63
  (*See also* ENERGY LOCK; Featherlite; KEEVA)
post-and-beam panel systems, 34, 50, 55–57
  (*See also* Amhome; ThermoFormed)
post-and-beam systems, 31–32, 33, 39, 44, 109, 111, 117, 122, 133, 204, 219–221, 262, 293
  spacing of members, 32, 44, 56–56, 101, 109, 188, 189, 205, 206
  (*See also* Amhome; ENERGY LOCK; Featherlite; KEEVA; post-and-beam block systems; post-and-beam panel systems; ThermoFormed)
projections (*see* spans and projections)

Quad-Lock, 51, 59, 60, 98, 112, 188

RASTRA, 50, 55, 98, 112, 187
R2000, 11

R-FORMS, 49, 55, 98, 112, 187
R-value:
  of foam, 67, 69–73, 76, 78, 79
  mass-corrected, 245–247, 250
  measures of, 8, 100, 241–242, 247
  of unit, 97, 105
  of walls, 7, 10–11, 87, 154, 242, 256
Reddi-Form, 52, 61, 62, 99, 106, 113, 188
reinforcement:
  of floors, 159–160
  installation, 261, 265, 266, 272–273, 290
  minimum, 207, 208, 231, 253, 254–255, 256
  schedule, 16, 200, 206, 223, 232–233, 254
  tensile strength, 206
  of walls, 17, 115, 117, 154, 179, 180, 189, 200, 203, 208, 210, 213–215, 217, 222–223, 235, 244–245, 256
REWARD, 52, 61, 99, 113, 188
roofs, 162–165, 243
running bond, 279

SBCCI (*see* Southern Building Code Congress International)
scaffolding, 47, 266, 269, 284–285
  (*See also* bracing/scaffolding systems)
SDI (*see* smoke developed index)
SmartBlock SF 10, 53, 62, 99, 113, 188
SmartBlock VWF, 52, 60, 99, 113, 188
smoke developed index, 16, 77, 78
Society of the Plastics Industry, 78
sound resistance, 22
sound transmission coefficient, 22
Southern Building Code Congress International, 98–99, 107, 203, 235
spans (*see* spans and projections)
spans and projections, 11, 13–15, 16, 128, 134–138, 180, 197–199
specialty unit, 40, 57, 59, 61, 98–99, 101, 105–106, 186
  (*See also* type of unit)
SPI (*see* Society of the Plastics Industry)
stack bond, 280, 286
standard unit, 32, 37, 40, 41, 44, 61, 186, 219
standards:
  structural, 16
STC (*see* sound transmission coefficient)
Steiner Tunnel test, 84–85

stops, 41, 46, 57, 60
strength of walls, 15–16, 17, 102, 124, 204
stretcher unit (*see* standard unit)
stucco, 13, 90–91, 171, 173, 176
  (*See also* cost, of exterior finish; finish, exterior)
successive blocks, 302
successive plates, 303
surface relief, 13
sway frames (*see* frames, sway)

termite shield (*see* insects, provisions for)
thermal mass, 7, 9–11, 102, 154, 243–247, 249, 256
thermal resistance (*see* R-value)
ThermoFormed, 50, 56–57, 98, 112, 187
Therm-O-Wall, 53, 61, 113, 188
tie ends, 37, 38, 49–53
tie heads (*see* tie ends)
ties, 31, 37, 42, 49–53, 59, 60, 61, 106, 283
  types of, 42–43, 45
top bracing, 266, 267
treatment gap (*see* insects, provisions for)

U-block (*see* lintel unit)
U-value, 241
unassembled Lite-Form [*see* Lite-Form (nonassembled)]
unit (*see* type of unit)
utility lines, 103, 162, 166–170, 266, 268, 270, 308–309

vertical cavities (*see* cavities)
vertical member (*see* post)

wall dimensions:
  gross exterior area, 4, 186
  thickness, 111–113, 135, 206
waste, 181, 192
water-to-cement ratio, 294
webs:
  concrete, 40
  foam, 37, 44, 98–99, 102–103
  structural, 215–217, 219–221
  tie, 37

wet set, 275–277
wind resistance, 17, 102, 124, 199
  provisions for, 86, 179, 180
windows, 243
  and moisture, 18, 87, 89–91, 146
  mounting, 146–151

windows (*Cont.*):
  (*See also* openings)

XPS (*see* foam, extruded polystyrene)

ABOUT THE AUTHORS

PIETER A. VANDERWERF is an assistant professor at the Boston University School of Management, and the author of McGraw-Hill's *The Portland Cement Association's Guide to Concrete Homebuilding Systems* and *The Insulating Concrete Forms Construction Manual*.

STEPHEN J. FEIGE is on the staff of Grassi Design Group of Boston, Massachusetts, has years of experience in building, remodeling, and designing homes.

PAULA CHAMMAS is a design engineer for Mercedes-Benz in Frankfurt, Germany, who has extensive experience in the modeling and analysis of diverse materials and products.

LIONEL A. LEMAY is program manager for residential technology at the Portland Cement Association in Skokie, Illinois.